*The Dark
Behind the Curtain*

The Dark
Behind the Curtain

Gillian Cross

Illustrated by
David Parkins

Oxford University Press
Oxford Toronto Melbourne

Oxford University Press, Walton Street, Oxford OX2 6DP

Oxford London
New York Toronto Melbourne Auckland
Kuala Lumpur Singapore Hong Kong Tokyo
Delhi Bombay Calcutta Madras Karachi
Nairobi Dar es Salaam Cape Town

and associated companies in
Beirut Berlin Ibadan Mexico City Nicosia

Oxford is a trade mark of Oxford University Press

British Library Cataloguing in Publication Data
Cross, Gillian
The dark behind the curtain.
I. Title
823'.914[F]   PZ7
ISBN 0-19-271457-0

Photoset in Great Britain by
Rowland Phototypesetting Limited, Bury St Edmunds, Suffolk
Printed and bound by
Biddles Limited, Guildford and King's Lynn

Chapter 1

The heavy red curtain swirled outwards and settled into long folds. From behind it shuffled a stooped figure, alert and menacing, his fingers gripping something long and straight which glittered as he stropped it restlessly against his other hand.

'They all come' – the voice was low and gloating – 'rich and poor, old and young, they all need the barber.' His feet inched forwards over the boards. 'I sit them in my chair and I lather 'em up to the nose, watching their silly faces in the mirror, dreaming of fine dinners and fancy women. No eyes for the barber as comes creeping, creeping . . . .'

'Then, all of a sudden –' the figure jerked upright and in his right hand something bright flashed once, slicing murderously down. '– I *cuts* 'em! From ear to ear! And the blood drips down their shirts and puddles on the floor.'

The left hand traced the slump of a head, the slow

sliding of a body. 'And then' – briskly – 'into the cellar with 'em and off to be chopped up and rolled in pie crust.'

His laugh shivered through the silence.

Jackus, leaning against the door at the back of the Hall, shuddered involuntarily. Somehow, he had not expected anything so good. As the laugh died away, there was a hush. For a full minute, nobody spoke.

Then fat Ann Ridley broke the tension with a nervous giggle:

'That was *beastly*, Marshall!'

'That was *tremendous*, Marshall!' said another voice. Miss Lampeter jumped to her feet, her fair hair swinging, and clapped her hands. 'That's just what I wanted.'

The figure on the stage smiled and thrust the metal ruler into a blazer pocket. 'Had you all shaking with terror, did I?' Marshall said jauntily. His eyes glanced over the rows of faces, flickering with amusement. Then he saw Jackus at the back of the Hall and his smile broadened. 'Even you, Jackanapes?'

'Take more than you to terrify me,' Jackus retorted calmly. But he did not feel calm. They had all turned to stare at him and someone started to whisper. He had planned to slip up to Miss Lampeter at the end and talk to her without being overheard. Now he was forced to walk up the Hall like some kind of public procession.

'I say, you fellows,' murmured Stephen in his Billy Bunter voice, 'look what's crawled out from under a stone.'

Ignoring him, Jackus paused by Miss Lampeter. 'Mr Garner sent me. He said you wanted me. For the play.'

Miss Lampeter smiled, consciously gentle. 'Yes. I'm pleased you've come, Colin. Now Alan Benning's let us down we need a new Jarvis Williams. We were in rather a pickle.'

She laid a hand on his arm, and Jackus had to make an effort not to shake it off. Before he could think of something flippant to say, Ann giggled again:

6

'We never thought you'd really come. Not to be in a *play*.'

Jackus sniffed. 'Just fancied seeing you all make idiots of yourselves.'

He felt them bristle. Little Benny Harris clenched his fists and went pink. 'We knew you'd be rotten. Marshall said –'

'I said,' Marshall sat down on the edge of the stage and swung his legs, 'that you would come. Just to annoy everyone.'

'Thanks a lot.' Jackus scowled at him.

'I think it's very good of Colin,' Miss Lampeter said hastily, a trifle too enthusiastic. 'I don't know what we'd have done without him. I want you all to make him feel welcome and help him catch up what he's missed.'

Ann picked that up at once, of course. Ever bossily helpful. 'You can have my script, Jackus. I've learnt all my words already. Here.'

She held out a bundle of papers and, without looking at her, Jackus closed his fingers reluctantly round it. After all, he had no choice.

'Try and read it through before Thursday,' Miss Lampeter said. 'That's when we're meeting again. It's not a rehearsal. More a sort of workshop. But it'll help if you know the play.'

Jackus grunted and let his arm swing, making no attempt to look at the script. For a moment Miss Lampeter paused, slightly puzzled, as if she expected him to say something. Then she clapped her hands briskly:

'Right, everyone. That'll do for today. We've done a lot of good work. See you on Thursday, in the History room.'

Gathering her papers up in an untidy bundle, hugged to her chest, she smiled round at them all and swung off towards the door, her spiky heels clipping sharply on the floorboards.

'Quite the little producer, isn't she?' Jackus muttered sarcastically.

7

Benny Harris was still annoyed. He whirled round and thrust his face at Jackus:

'You shut up! *We* know you've only come to make trouble, even if she doesn't. And we're enjoying dóing this play. You dare try and mess it up.'

'Dear, dear.' Jackus grinned irritatingly. 'What's got you so hot under the collar? Soft on her, are you?'

They were all glaring at him now. It gave him a sort of sour pleasure to bait Benny, as a way of getting his own back. But he did not intend to fight. When he saw Benny's fists double up, he side-stepped, ready to pass the whole thing off with a joke. But, as he glanced up, he suddenly saw Marshali smile scornfully and look away, refusing to meet his eyes. In a spasm of rage, he lashed out at Benny's silly, fair face.

The next moment the two of them were rolling on the floor, battering at each other, while Ann said feebly, 'Oh, do *stop* it. Don't be stupid.'

Pulling a rude face at her, Jackus rolled Benny on to his back, scrambling to sit astride his chest.

'Feeble as well as a creep, aren't you, Harris?' he muttered triumphantly and drew back his fist. But at that instant, he was aware of an odd silence. Marshall's legs stopped swinging, abruptly.

'Oh, crikey! Oh lor!' murmured Stephen.

'Colin Jackus!' rapped a voice from the back of the Hall. 'What are you doing?'

Jackus groaned and looked over his shoulder. Old Garner was striding up from the door, his bald head shiny and his face sharp with anger.

'Well?' he said, standing over them.

Feeling foolish, Jackus jumped up and Benny got sulkily to his feet.

'He started it, sir.'

'Well, Jackus?'

Jackus looked at the ground. 'I did hit him first.' No point in trying to explain.

Mr Garner nodded sarcastically. 'It must have been a

great challenge to you. A boy of Benny's size. Wait here, Jackus. I'll speak to you when the others have gone.'

Jackus stood awkwardly while the rest of the cast edged by Mr Garner, turning their eyes away in embarrassment. Only Marshall gave him a long stare as he passed.

'Not a very good beginning,' Mr Garner said, when the Hall was clear. 'I thought you had more sense.'

Jackus shrugged. 'Don't want me in the play, do they?' he said sullenly. 'I knew it would be like that.'

'So you made sure of it by fighting Benny?' Old Garner sighed and sat down. 'I would remind you that we made a bargain. A bargain it is in your interests to keep.'

'I did keep it,' Jackus said quickly. He waved his script under the headmaster's nose. 'See? I told Miss Lampeter I'd be in her rotten – in her play.'

Mr Garner looked at him sharply. 'Don't pretend to be more stupid than you are. You know that I didn't mean you to do it grudgingly. What I want from you, Jackus, is co-operation.'

Jackus said nothing. Old Garner watched him for a second and then leaned back in his chair.

'I don't want you to think I've gone soft in my old age. I didn't send you down here as a way of letting you off, you know. I was quite aware that it would be difficult. But I was prepared to take a chance on you.' His face wrinkled fastidiously. 'I have a particular dislike of involving the police in school affairs. Even where theft is concerned. But if you break our bargain –'

'I know,' Jackus said wearily. 'You'll call in the law. You told me.'

'I'm glad we understand each other so well,' Old Garner murmured wryly. 'And I don't expect to hear that you have been causing any more trouble.' He contemplated Jackus's rebellious face and chuckled. 'You never know. If you practise being pleasant, you may find that it makes people more friendly towards you.'

'Yes, sir. Can I go now, sir?'

The headmaster nodded and Jackus slouched off down the Hall, in an evil temper. As the door closed behind him, he aimed a kick at the wall, knocking yet another flake of paint from its battered surface. 'Bloody play!' he snarled. 'Bloody Miss Lampeter! And bloody – Old – Garner!'

But it did not make him feel any better. Still furious, he snatched his jacket from its peg in the empty cloakroom and stamped off down the school drive.

When he turned off the road, down the footpath which led to the canal, he saw a tall shape at the far end. It was leaning against the corner of a building, its hands in its pockets and its lean shape reflected vaguely in the scummy surface of the water. Not altogether surprised, Jackus took his time walking down the path. As he drew level, he nodded.

'Hi, Colin,' said Marshall lightly.

'Hi, Colin,' Jackus said. 'Waiting for someone?'

Marshall inclined his head. 'Just thought I'd find out how things went. Did Old Garner give you a bad time?'

'Not really. Just told me what a wicked boy I was to thump poor defenceless little Benny Creep-Face.'

'Not *that*.' Marshall flicked his fingers impatiently. 'Before. When you went up to his office.' He was watching Jackus surreptitiously, through his eyelashes. 'I thought you'd be in handcuffs next time I saw you.'

'Planning a dramatic rescue, were you? Hijacking the Black Maria and driving me to safety?'

'Of course.' Marshall grinned. 'Now come on. Give. What did he say to you?'

'You could have been there yourself, couldn't you?' Jackus said gruffly. 'If you're so keen to know.'

'Ah!' Striking a dramatic pose, Marshall clapped a hand to his forehead. 'A flash of temper! Can it be that the worm has turned?'

Suddenly, Jackus realized what it was all about. He punched Marshall amiably. 'Thought I'd got myself off by Telling All, did you? I bet that had you worried.'

'Worried? Me?' Marshall raised his eyebrows. 'Why should I be worried?'

They began to walk along the tow-path, swinging their bags. On either side, tall warehouses stretched up, shutting out the feeble evening light. As he went, Jackus kicked sideways at little stones and they fell into the water with thick, oozy splashes.

'No handcuffs,' he said at last. 'Just a lecture on my idiocy and the way I never act as if I were part of the school. Usual Old Garner rubbish.' He laughed lightly. Somehow, he did not want to tell Marshall how unpleasant it had been. 'And now I'm in the play. Part of the cast of *Sweeney Todd, the Demon Barber of Fleet Street.*' He swept a bow, grandly theatrical. 'The start of a new and sensational career.'

Marshall gave him a sly, sidelong glance. 'You won't like it, you know,' he said abruptly. 'It's not your sort of thing at all.'

'You're telling me.' Jackus kicked bitterly at the towpath, knocking a hole in the ground. 'I'm not like you. Don't go out of my way to make an exhibition of myself.'

'I'm worth an exhibition,' Marshall said grandly. He flung his bag into the air, catching it neatly just as it looked ready to fall into the water. 'I'm going to give a stunning performance. I'll have the audience sweating with nightmares after they've seen me.'

'You do that to people anyway,' said Jackus. 'Without trying.' He sniffed. 'Seems a pretty stupid sort of play to me, anyway. All about some old geezer knocking people off and putting them in pies? What rubbish.'

'Now, now.' Marshall wagged a finger at him. 'You'd better not let the Lamppost hear you talking like that. She's all set to make this the best Lower School play ever. Prove she can do better than Old Garner with his everlasting Shakespeares. She won't like it if you knock it.'

11

'Have to lump it then, won't she?' Jackus growled. 'If she's so keen to have me in it.'

'Oh, I expect she thinks she's going to save you,' Marshall said flippantly. 'Turn you into a respectable member of the community. Like me.'

'Ha ha. Very funny.' Suddenly Jackus felt the urge to tease Marshall. 'I could shake that virtuous image of yours if I wanted to, couldn't I?'

Marshall did not bridle as anyone else would have done. Instead, he smiled. A tight little smile. 'I think you ought to keep out of my hair,' he said dangerously. 'Tell them you don't want to be in the play after all. Because you don't really, do you?'

Leaning forward with a swift, darting movement, he snatched the rolled up script from under Jackus's arm and held it out over the water of the canal, letting the pages flap slowly in the wind.

'Go on, Jacko. Why don't you tell her what she can do with her precious script?'

As the breeze ruffled the surface of the water, the reflection of the paper rippled and fragmented.

'Go on,' Marshall said again, softly. 'Shall I drop it in?'

In his mind's eye, Jackus saw the white pages spill out onto the scum, spreading wide, sucking up the dirty water. They would grow soggy in the end and sink down into the mud and slime at the bottom, mercifully lost.

'Funny joke,' he said heavily. 'Give it back, Marshall.'

'Oh, go *on*. Just one little splash and it'd all be over. You wouldn't have to get up on the stage and make a fool of yourself.' His voice was humorous, but Jackus, watching him, saw the calculating look in his narrowed eyes.

'You're serious, aren't you?' he said slowly. 'You really don't want me in the play with you.'

Marshall's face, pale in the twilight, twisted into an unconvincing grin. 'Of course I want to have you in it.

12

You're my best mate, aren't you? Auntie Mary's little boy. Friend of my childhood.' He laughed quietly. 'But you would like me to chuck this in the canal. Splash!' He pretended to drop it and it almost fell from his fingers. As he tightened his grip round it, he grinned.

Jackus suddenly tired of the game. Because there was nothing he would have liked better than to see the script fall into the water. 'Pack it in, Marshall,' he grunted. 'It's not funny. I've *got* to be in the rotten play. Old Garner said. Otherwise he'll call in the police.'

'Oh, I *see*.' Marshall's face cleared suddenly, understanding. 'So that's the bargain. That's how you got yourself off. By promising to be a good, helpful little angel.'

With a crow of laughter, he waved the script frantically over the water and Jackus, afraid that he might actually drop it by mistake, reached out to snatch it. He slipped, and his foot slid neatly into the canal. He saved himself from following it only by falling to one knee and clutching at the concrete edging. From the black, oily depths which he had disturbed came a stench of decaying rubbish.

'Now look what you've done!'

Marshall held his nose and dropped the script delicately on the ground. 'Let's hope that's all you're going to stir up.'

Cursing, Jackus dragged his leg out of the water and pulled off his shoe to empty it. 'You're an idiot. Of course I won't tell anyone. Anything.'

'What a pal,' Marshall said politely, turning away. 'See you at rehearsals then.'

He walked off down the tow-path. 'I still think it's a barmy play,' Jackus shouted after him. But Marshall had already rounded the corner of a warehouse and disappeared. Jackus was left alone in the half light between the high walls with a soggy trouser-leg and the stink of decay filling his nostrils.

*   *   *

13

'Hallo, Col,' Mrs Jackus said as he came in. 'Have a good day?'

No. I was threatened with the police, I had a fight with Benny Harris, and your blue-eyed boy pushed me in the canal. He could just imagine how her face would crumple with disappointment if he said that. Instead, he shrugged.

'They've put me in the Lower School Play.'

'Oh, Col!' She beamed at him. 'That's lovely. Who else is in it? Big Colin?'

'I wish you wouldn't call him that,' Jackus said irritably. 'Marshall. His name's Marshall.'

'He'll always be Big Colin to me.' Mrs Jackus smiled sentimentally. 'Ever since that first day when Rose brought him round, just after you were both born. I'll never forget her face when we realized each of us had chosen the same name.'

'I know,' Jackus said wearily. 'And you vowed the two of us would be best friends, like you two. You've told me a million times before.'

His mother looked hurt. 'Well, we were right, weren't we? You *are* friends?'

Silly question, thought Jackus. It was like asking if you liked your feet. Marshall was Marshall. Who'd always been there. Aloud, he said, 'I dunno. He's O.K. Bit of a Clever Dick.'

'It wouldn't hurt you to be a bit more of a Clever Dick,' Mrs Jackus said, rather sharply. 'Sometimes I think –'

'I know, I know,' Jackus interrupted rudely. 'You'd like me to be just like him. You've been waiting for it ever since we were babies. Well, you can just go on waiting!'

Crossly, he stamped into the living-room and sat down. So that she couldn't talk to him any more, he began to read the nearest thing, which was the script of *Sweeney Todd* that he held in his hand.

Extract from the Diary of Ann Ridley
Tuesday, 10th November

'. . . It started off as a horrible rehearsal. Whatever we did, the Lamppost got in a temper. She stalked up and down, saying "You haven't got the right idea, any of you. You still think it's a funny play. I want you to see it's *frightening*. That's why I rewrote the script."

The more we tried, the worse she got. I'd learnt all my words, on purpose, but she didn't even seem to notice. Every time I spoke, I could see her scribbling things in that little notebook of hers. It made me feel really awkward. Like a sack of potatoes.

It was Marshall who saved the day. Right at the end, she told him to do his opening speech – and he was *marvellous*. He shuffled about like a creepy old man and talked in such a beastly voice that I went all goose-pimply. It was a real bit of acting. I don't know how anyone so super (!!!) could be so like a dirty old man. The Lamppost was so excited that she could hardly sit still.

Everything would have been lovely except that, after that, Colin Jackus turned up. It seems he's going to play Jarvis Williams. That means I've got to act with him. (Yuck!) He looked at us all as if he wanted to kill us, and kept making stupid jokes. I hope he doesn't spoil it all.

Then he started fighting Benny Harris. Poor little Benny. I don't see how anyone could hit a little boy like him. And Mr Garner came in, so there was a really horrid row. It quite upset me.

So it would have been a completely foul day if it hadn't been for Marshall.'

15

Chapter 2

Jackus loitered outside the door of the History room,
running a grubby finger over the moulding of the door
frame.

From inside, he could hear Miss Lampeter's high-
pitched, eager voice. 'You must look more tired, Ann.
Remember, this is the third shirt you've sewn today and
your eyesight's going. . . . That's right, Benny. You're
shaking with fear. You've got to climb the chimney with
your brushes, and you know that boys often get stuck
up there and die. And they've threatened to light a fire
underneath you to make you climb higher.'

Slowly Jackus pushed the door open. Inside, everything looked chaotic. The blackout curtains were drawn,
as if for a film, and the desks were pushed back, leaving
a large space in the centre of the room. People were
moving about silently, miming incomprehensible
actions. He could see Ann, sitting cross-legged on the
floor, drooping forwards as she pretended to sew, and

Benny, scrabbling his arms and legs in an unconvincing imitation of climbing, with an expression of terror on his face. Everyone looked cowed and weary, except for Marshall. He was strolling about, with his usual superior smile, peering scornfully at the huddled shapes.

Miss Lampeter caught sight of Jackus and darted up to him immediately. 'Ah – Colin.' He was braced for a scolding, knowing that he was half an hour late, but she did not even mention that. She just gave him a sickening smile and patted his hand. 'Nice to see you. I expect you wonder what on earth's going on?'

'Looks like playtime at a lunatic asylum.'

She laughed brightly. 'You won't feel like that once you get involved. It's not meant to be a performance. It's just an idea of mine to help everyone get into the right mood for the play.'

She seized his arm and dragged him across to a corner. 'They're all acting out different things, you see, pretending to be children in the Victorian slums. Ann's trying to make her living sewing at tuppence a shirt. Stephen's minding machines in a factory, although he's so tired he can hardly keep awake, and – oh, I haven't got time to explain all the things to you, but the point is that they're all *victims* – just as they are in the play.'

'Marshall doesn't look much like a victim,' Jackus said, watching him lean over Ann, stabbing with an impatient finger, as if he were making a complaint.

'Of course not. That's the point,' trilled Miss Lampeter. 'He's the person who's exploiting them all, the one they're all afraid of. Just like the play again, you see.' She surveyed Jackus thoughtfully. 'Now, what am I going to do with you? Let me see. . . .'

He wanted to say that he thought the whole thing was stupid, and that he had no intention of prancing around like a loony, but he did not dare.

'I know.' With a quick gesture, she pulled the yellow

17

silk scarf from round her neck. 'You can be a blind beggar. How about that? Hang on while I tie this over your eyes.'

Before he could even frown, she had knotted the scarf round his head, blocking out the sight of Marshall's amused grin.

'There. Now you're blind from birth. And starving. Your only hope of a meal is to cadge a few pennies from passers-by. Squat down and let's hear you beg.'

Feeling an idiot, Jackus crouched on the floor. He could not see anything at all. From the scarf came a faint, incongruous scent of honeysuckle.

'Go on,' Miss Lampeter said impatiently.

It was like some mad party game. The kind where you stumble about in the dark while everyone else giggles because you're making a fool of yourself. 'Give us a penny, miss,' he whined awkwardly. 'Penny for the poor blind beggar.'

'That's it,' he heard her chirrup. 'Splendid. Keep it up.'

Then her heels clicked on the boards and she was gone. He was alone in the dark, with feet thundering past him and no idea who was near. It felt as though someone was sure to kick him by mistake as he huddled there on the floor. In self defence, he kept up his whine.

'Blind, blind. Give us a penny. Penny, kind sir. Blind, blind.'

There was a rustle of clothes close to his ear and he heard someone breathe, stooped close beside him.

'There you are, poor boy,' said a voice. 'It's not much, but it's all I've got.'

No mistaking that smug voice. It was Ann. Poor shirtmaker indeed! She sounded more like someone giving in her homework. Jackus fumbled forward and pinched her solid calf in its thick woollen stocking.

'Yes, I can tell you're nothing but skin and bone,' he said cruelly.

18

'You beast! I'll have my penny back, for that.' She stooped again and then he heard her flounce off, leaving him on his own.

His stomach rumbled juicily. It was hours since lunch, and he was starving. Good thing he did not really have to beg enough pennies for his next meal, he thought wryly. He would be dead before he had scrounged the price of a cup of tea. As the feet padded past, he reached out, trying to clutch someone.

'Blind, blind. Penny for the blind man,' he called.

No one stopped. They were all busy with what they were doing, all trying to scrape together a living for themselves.

'Just for a loaf of bread,' he wailed, lunging and snatching as bodies brushed by him. Then his fingers closed on something. The edge of a trouser-leg. He gripped it tightly. 'Please, sir, give us something. Just so I can get a bit of food.'

'Let go,' said a cold, haughty voice. 'I don't believe in giving money to beggars. It encourages pauperism.'

It was Marshall, of course, playing the part he had been told to play. To his surprise, Jackus found that he was really angry. He lashed out with his other hand, catching a hard ankle.

'Dear, dear,' Marshall's ordinary voice said softly. 'Getting quite carried away, aren't you?'

Then Jackus felt the trouser-leg twitch away from his hand and he was left fuming.

'You'd see your own mother starving, you would,' he muttered. But he felt amazed at himself. It really got you, this stupid game. Made you feel defenceless and unable to help yourself. Or was it just that he could not see anything? He would have liked to rip the blindfold off.

Suddenly, everyone shrieked. Behind the scarf, Jackus was aware of a deeper blackness. Quickly he reached for the knot and undid it. For one crazy moment, he thought that he really had gone blind. Then he heard Miss Lampeter laugh, and he realized that she had switched off the light.

'Time for the next exercise,' she said. 'Now you're all nicely loosened up.'

In the darkness, the shrieks stopped, and the figures stood like black humps, their faces indistinguishable, waiting for something to happen. There was the scrape and hiss of a match and a candle flame leaped brightly. Miss Lampeter was standing by the door, her face sharpened by the single, flickering light. When she spoke again, her voice was lower, crooning, as if she were trying to lull them into a dreamlike state.

'Now, I want you all sitting in a circle on the floor. Everyone except Marshall.'

Shuffling and pushing, they arranged themselves. Jackus found himself next to Ann, her solid shoulder making a dark, shapeless mass in the corner of his eye. Miss Lampeter stepped carefully into the circle, bent down, and placed the candle on the floor.

'Now,' she murmured, still in the same soothing voice, 'it's different. I want you all to imagine that you're child thieves.'

Behind him, Jackus heard a quick splutter of laughter from Marshall and he pressed his lips together, annoyed.

'You're still poor,' Miss Lampeter went on, 'but now you're corrupt, as well. You've been sneaking about all day, twitching handkerchiefs out of people's pockets and cutting ribbons to steal pocket watches. You're tired, and very hungry. And you're showing each other what you've got.'

She stepped out of the circle in one long-legged stride and the candle flame dipped and sputtered as she passed, throwing long shadows up the faces which bent around it, turning them sly and conspiratorial. Fingers darted, sending spikes of black across the floor, turning imaginary loot. Someone chuckled greedily and Benny Harris leaned forward, his choirboy's face suddenly cunning and mean in the jerky gleam of the flame.

Ann turned her head to look at Jackus. 'What did you pinch, then?' she whispered.

'Oh, shut up!' Jackus growled. 'You're as bad as the Lamppost.'

Ann's face crumpled as if she were hurt and she bounced round to look at her other neighbour, leaving Jackus facing the wall of her back.

Miss Lampeter had started to talk again, her voice so low that it was only just audible. 'Evil. Can you feel it? You've all grown up in an atmosphere of evil, breathing it in, absorbing it through the very pores of your skin. Somewhere, some part of you knows that this is not the way that people ought to live, but you can't imagine anything else.'

Inside his head, Jackus snorted, but somehow the sound did not emerge. It was not easy to be as scornful as he wanted to be. For there was certainly something strange in the air. A feeling almost of wretchedness, of defeat. His skin prickled and he found suddenly that he was genuinely cold, as though the temperature in the room had dropped several degrees.

Rebelliously, fighting the feeling, he plunged his hands into his pockets, turning over the prosaic, ordinary shapes of the things inside – coins, conkers, a rubber band.

Behind him, he could hear a low, steady whisper. Miss Lampeter was explaining to Marshall what she wanted him to do. All at once, her voice lifted slightly, so that her words floated across the circle.

'You're waiting,' she hissed. 'Waiting for the man who will take what you have stolen and give you some food. You don't know his name. You all call him Old Stony, and you're scared of him. He'll want everything that you've taken. If he catches you trying to cheat him, he'll have you beaten up. And now he's coming.'

Benny's fair hair gleamed suddenly in the candlelight as he glanced nervously over his shoulder. There was a

21

rustle of air, as people caught their breath, and the candle flame jumped and sagged, shrinking the pool of light and letting the shadows from outside eat into the little patch of brightness.

From the far side of the room came the sound of dragging footsteps, slow and deliberate, like an old man climbing stairs. The conkers under Jackus's fingers were hard and smooth, but they had stopped being reassuring. Ridiculously, he found himself beginning to grow afraid. *It's the dark*, he said inside his head, but it was no use. The coins in his pocket clinked together and he winced at the sound. Secret money, kept back from the pool. Was he going to get away with it, or would fists thud into his face, boots crash against his spine?

The feeling struck a nasty echo at the back of his mind. He had been in the same situation before – crouching in the dark, afraid that he would be caught with stolen property. To dispel the shudder that began to creep slowly up his body, he glanced sideways at Ann. There could be nothing more down-to-earth and reassuring than the sight of her ungainly bulk.

But for a moment he could not see her. It was as if there were a thicker darkness between them, shutting off any dim flickers of light. Vaguely, he thought that someone must have moved across the circle. But he had not been aware of any movement. Then the darkness cleared. She looked over her shoulder and frowned at him.

The dragging footsteps drew nearer. Marshall stepped over bodies into the circle and looked down at them all, the candle-flame sending shadows the wrong way up his face, giving him devil's eyebrows. Under his arm he held a loaf of bread, and there was a quick hiss from around the circle as the huddled figures caught sight of it. But he held it up, well clear, as he scanned the ground in front of him, stretching out a toe to turn invisible heaps of loot.

All at once, Jackus was aware of something else beside him. An unpleasant smell of old sweat, rancid bacon and a curious mustiness, like the smell of jumble. Ann? He glanced round at her in surprise, remembering her clean, scrubbed face and shining hair. Could she really be smelling like that?

It grew stronger, making him want to retch, and he was just about to stretch out a hand and push her away from him when a blast of warm, stinking air caught him in the face, as though someone had turned towards him, breathing hard. A breath of onions and vinegar and bad teeth.

He spluttered slightly and, at that moment, Marshall spoke, grudging and harsh.

'Not enough. You'll have to do better tomorrow, or there'll be no food for you.'

With a scornful gesture, he tossed the loaf of bread down on to the ground, flicking crumbs from his fingers. At once, bodies launched forwards, hands reached desperately for the bread. Jackus felt a hard, bare arm push against his face, knocking him sideways, so that he sprawled on the floor. He gave a loud, outraged yell.

The next moment, the light flashed on and Miss Lampeter was at the edge of the circle, smiling down at them. The crouched bodies were motionless, caught in the middle of the frantic scrabble, and faces looked up, blinking in the glare. The loaf had been torn into a scatter of crumbling pieces. Benny had one jammed into his mouth and another clutched tightly in his hand. Beside Jackus, sprawled the other way, Ann lay spread-eagled on the floor.

'Oh, that was *very* good!' Miss Lampeter clapped her hands. 'You really felt it, didn't you? The hunger? The desperation?'

For a moment, no one spoke. Then Ann broke the uncomfortable silence. She sat up, rubbing her elbow.

'You didn't have to push me so hard, Jackus. I'll have a terrible bruise tomorrow.'

23

Miss Lampeter laughed. 'Got carried away, did you, Colin? Never mind.'

'But I didn't,' Jackus said slowly. 'I never laid a finger on her.'

Ann exploded.'Of course you did! Who else could it have been? There was no one else on that side of me. Anyway, I don't push people.'

Stephen nodded sagely. 'The esteemed and ridiculous Ann has never been known to use force.'

Jackus scowled. 'Think I'd lay a finger on *her*? I wasn't playing games like the rest of you. I thought the whole thing was stupid.'

'You think everything's stupid,' Benny burst out. 'I just wish you'd decide breathing's stupid. Then you might give it up.'

'That's *enough*!' Miss Lampeter said forcefully. 'I don't want to hear another word. You're *all* being stupid.'

'What do you expect from a crowd of Victorian paupers?' murmured Marshall. He stretched out a long arm and picked up one of the cleaner pieces of bread, that no one had chewed. Grinning round at them all, he started to eat it.

'Eugh, *Marshall!*' Ann said. 'I don't know how you can.'

'What's the matter?' Marshall drawled, amused. 'It's very good bread. Why shouldn't I eat it?'

'It's – you know – oh, it's been on the floor,' Ann said hastily. But she looked away from him, as though she had not given the real reason.

Miss Lampeter had begun to stack her books together. 'Time you were all off home,' she said, 'or your mothers will think I've kidnapped you.'

Ignoring the others, Jackus walked across the room to pick up his bag. He was puzzled. Why should Ann have made that fuss about being pushed when she had done the pushing. Hadn't she? He looked across at her. She was pulling on her coat, over her long-sleeved school jersey. All at once, he remembered the feel of the bare arm that had pushed him. The skin had brushed against his face, cool and naked. He felt a strange quiver run down the side where that rough arm had thrust at him. Then he shook himself back into common sense. Of *course* it had been her who pushed him. It must have been.

Chapter 3

'Colin.'

Jackus, diving out of the class-room door, stopped automatically at the sound of his name and groaned as he turned. He should have guessed it was the Lamppost. She was hurrying down the corridor towards him.

'Glad I've caught you,' she said breathlessly. 'I want you and Ann in the Geography room before lunch. To go through your first two scenes. All right?'

'Suppose so.'

'I thought you'd like to do your first bit of acting without everyone gawping at you.' *See how understanding I am*, said her smile.

Jackus grimaced. 'You mean, you don't want everyone to see me make an idiot of myself.'

'Why should you make an idiot of yourself?' she said, with an edge to her voice.

''Course I will. I never do any acting.' He snorted. 'It's a stupid idea to put me in a play.'

She stared steadily at him for a moment, her eyes cold. 'Look,' she said at last, 'I wasn't keen on it either, if you want to know. I'm putting a lot of effort into this production, and I don't want it ruined by a silly boy who's cross because he got caught out with his hands full of things that didn't belong to him. I only agreed to have you because Mr Garner asked me most particularly.'

So it had been Old Garner's idea in the beginning, had it? Jackus was surprised. Somehow, he had assumed that she had been the one to dream it up.

'We're lumbered with each other, aren't we?' she went on shortly. 'So we'll have to make the best of a bad job. But don't think I'll put up with any nonsense from you, because I won't.'

He was still looking sullen, and her voice changed, wheedling. 'There's no reason why you shouldn't enjoy it, you know. You haven't got a big part, but it's a good one. You're really the hero, if you want to see it like that. Have you read the play?'

'Took a look at it.'

'Well, then, you know how you find out what Sweeney Todd's up to, and how you get rid of him with his own chair. Don't you like the idea?'

'Don't mind tipping Marshall down a hole,' Jackus said, faintly amused.

Miss Lampeter frowned at him. 'You don't get on very well, you two, do you?'

Jackus glanced quickly at her, wondering what she meant. 'Marshall's O.K.,' he muttered. 'Just so toffee-nosed he could set up as a sweet factory.'

'Well, I'd be very sorry if –' Miss Lampeter paused for a second '– if anything between you interfered with the play. We all need to work together. Trusting each other.' She laid a hand on his arm and looked earnestly into his eyes.

'It's not me that doesn't trust people,' Jackus snarled. He twitched his arm away, so roughly that he knocked her. From behind, a heavier hand fell on his shoulder.

'Careful, boy,' a voice grated. 'Are you having trouble with him, Miss Lampeter?'

'Oh, no.' She said it too quickly, too emphatically. 'Colin is just fine. Aren't you, Colin? We were talking about the play, Mr Garner.'

'Ah, the play,' said Old Garner vaguely. 'Your melodrama.' He turned to Jackus. 'Miss Lampeter had a feeling you'd enjoy something like that more than Shakespeare. How's it going?'

'It's splendid,' Miss Lampeter said, before Jackus could speak. 'Isn't it, Colin?'

Jackus looked from one to the other. There was the Lamppost, keen to show she was putting on the Great Production of All Time. And Old Garner, sour-faced and quizzical, as if he thought they all needed a good dose of *Hamlet*. He compromised.

'O.K., I suppose.'

'Well, I'll see you at lunch-time, then.' Miss Lampeter nodded brightly at him and swung off up the corridor. Old Garner lingered.

'I won't pretend I'm entirely happy about you, Jackus,' he murmured. 'I've got my eye on you, boy. Mind you play the game with Miss Lampeter. She's taking a risk, giving you this part. And it is her first production.'

'Ah, but *she's* a fantastic producer,' Jackus said, verging on rudeness.

Old Garner was too fly to fall for that one. He smiled wryly. 'Get along, then.'

'No, no, Ann,' Miss Lampeter said, exasperated. 'You're not getting it at all. You've got to frighten Colin.'

'She scares me stiff,' Jackus said helpfully. Ann aimed a hefty punch at him.

'Stop it, you two.' Miss Lampeter was getting slightly rattled. A long lock of hair had fallen across her face and she brushed it back impatiently. 'There's not much time left, and I want to get these scenes right before you have to go off for your lunch.'

Ann's face wrinkled in distress. 'I am trying. Really I am.'

'Very,' Jackus murmured automatically.

Ignoring him, Miss Lampeter took Ann's elbow. 'Look, try to imagine it. You make your living by cooking pies with bits of people in them. So you're cold and cruel. But you must have a baker to cook for you. And you're frightened. Frightened that he'll give you away. You can't stop, though, because you're frightened of Sweeney Todd, as well. All that makes you crueller. And here comes this poor, skinny boy, with no friends or relations to worry about him. Just what you need. You think – hmm.'

'Hmm.' Ann looked at Jackus, narrowing her eyes. He lounged sideways on a desk and stuck his tongue out at her behind Miss Lampeter's back.

'You tell him that he must come down to your cellar,' Miss Lampeter went on, 'and he can't leave, ever, until he quits the job. And you know what you mean by that. You mean that you'll have him murdered. So you must be threatening. You sound like a landlady showing someone a bedsitter.'

'It's so difficult to feel creepy in school uniform in the middle of a geography room,' Ann said helplessly.

'You're *very* creepy in school uniform,' said Jackus. 'Just like a prison wardress.'

'*Please*, Colin.' Miss Lampeter was getting really irritated. 'We must get going. You'll have to go and have your dinners soon.'

'About time,' Jackus said. 'I'm starving.'

'Good,' said Miss Lampeter heartlessly. 'So you should be. Starving and desperate for a job. Go on, Ann. Get started.'

Ann put her hands on her hips and looked Jackus scornfully up and down. 'Be off with you!' Her voice came out as a deep snarl. 'I can't be bothered with beggars.'

'That's better,' breathed Miss Lampeter.

'I'm not a beggar, mum,' whined Jackus. He had meant to put on the voice he had used when he was pretending to be blind, but somehow it came out differently. Cheekier. He saw Miss Lampeter sit up, and he emphasized the cheekiness. 'I'm just a poor lad as is looking for an honest job of work.'

Suddenly, he got the idea. He was chancing his luck. Chatting up this fat old bag so that he could get some money out of her. Because without a job, he would starve. In spite of his anger at being there, he began to be interested. He grinned chirpily at Ann. 'Have you got a job then, mum?'

Although he had his finger in the script, he found that he knew the words already. That was good, because he needed to keep his eyes on Ann. See how she was taking it, while he went on about what a good fellow he was.

And it was because he was watching her so closely that he noticed the change in her. To begin with, she was wooden and lumpish. Fat Ann Ridley, in her school uniform, trying to be sinister. She was so bad that they plodded through the first scene rather slowly.

Then, all of a sudden, she altered. They had swept straight on into the second scene, where she took him down into the cellar to show him his duties. As she pretended to point out the ovens and explain the routine of making up the pies, an unpleasant, crafty look slid over her face. Her eyes shifted evasively, and she kept licking at her bottom lip. On her forehead, small beads of sweat glistened damply.

'There is one thing you must understand if you are going to work for me,' she said. And her voice was odd, fat and evil. 'You may not leave this cellar. Not ever. Not for any reason.'

Ridiculously, Jackus felt a thump of panic. He had to

30

work hard to get the bravado back into his voice as he answered. 'But I can leave the job, can't I? Whenever I choose?'

'*Whenever* you choose.' Ann smiled chillingly. 'I won't keep you if you get restless. Never fear. I shall see that you go – to where you can be most useful.' She chuckled. 'Now I shall leave you to study your duties. But I shall be back.'

She waddled off, heavy-footed, and suddenly Jackus's concentration snapped. He stopped being Jarvis Williams, the cheeky, starving Victorian boy, and went back to being Colin Jackus, behaving like a loony in the Geography room. He fumbled, stuttered, and forgot his next words.

'Oh, do go on!' Miss Lampeter urged. 'You're doing fine. That was – what's the matter?'

'Dunno.' He blinked, vaguely, and looked at Ann as though there were something odd about her face. 'I was – I dunno.'

'You were *acting*.' Miss Lampeter swung her legs and grinned. 'You quite surprised me, both of you. I didn't know you had it in you.'

Ann was frowning. 'I think we should stop now,' she said abruptly. 'I want my lunch.'

She stood there, fat and obstinate and unhappy, and after a moment Miss Lampeter gave a confused nod. 'All right. I think you've had enough. You've both done really well. You know,' she laughed, 'I'm beginning to feel really excited about this play.'

As she slipped out of the door, Ann gave a long shudder and looked across at Jackus. 'I'm not,' she said in a strange voice. 'I think it's a horrid play. And Mrs Lovett's a horrid woman. I wish I didn't have to be her.'

Jackus was too surprised to be rude. 'But you were really good,' he said slowly. 'In that second scene, I could actually imagine you were her.'

'Do you know what I was thinking?' Ann picked up

the board rubber and banged it absently against a desk, sending clouds of chalk dust flying. 'All the time I was talking? It was like a voice in my head, saying, *That one won't make much of a pie. All skin and bone.* On and on and on. And it was like my voice – and yet it wasn't like my voice.' She looked nervously at him. 'Go on. Tell me I'm stupid. That's what you're thinking, isn't it?'

'No it isn't,' Jackus said slowly. 'What I was thinking was that you weren't like yourself when you were talking either. It was peculiar. I can't really explain.'

He studied her face. It was pale and shaky, her bottom lip trembling. Half of him was thinking that it was ridiculous of her to make such a stupid fuss about acting a bit of a nonsensical play. But the other half understood, reluctantly, that she had had some kind of shock, that she could not properly explain.

'Look,' he said suddenly, 'you have sandwiches too, don't you? Why don't we both eat them up here? Have a bit of peace and quiet?'

He could tell she understood. Normally that would have made her giggle furiously. Instead, she gave him a grateful smile.

'O.K. Hang on. I'll just get mine.'

Jackus opened his bag and pulled out the sandwich box. He had just picked up a cheese and tomato sandwich when he heard Ann scream. 'You beast!' she shouted.'You foul, horrible *beast!*'

His head jerked up. She had gone purple in the face, as if she were choking. For a moment he did not understand. Then he looked down. At her feet lay the whole contents of her sandwich box, tumbled out on the floor. Four little ham sandwiches, an apple and a Mars bar. And every one had a bite taken out of it. The semi-circular marks of teeth were quite plain. In the sandwich box, which she still held in her hand, were the mouthfuls that had been bitten out, sticking to the side of the box, gooey with saliva. As he watched, she shuddered and dropped the box on to the floor with a thump.

'That's why you wanted me to stay.' She gulped. 'Just so you could see my face when I – eugh!'

She charged at him, her fists pounding the air. Dropping his own sandwiches, Jackus grabbed her wrists and tried to reason with her.

'Don't be daft. I didn't do that. Why does it have to be me?'

'They were all right when I came in here. I'd only just got the Mars bar from my pocket to put it in. So there's only you that could have done it.'

Unable to hit him, she was panting close to his face, her arms wriggling with frustrated rage between his fingers. Jackus wondered for a moment whether she was teasing him. But he could feel her tremble. She was really upset. He tried to calm her down by speaking slowly and rationally.

'Look, Ann, I don't know who did that, but it wasn't me? How could it have been? You were here all the time.'

'Oh sure,' she gasped. 'It was Miss Lampeter, wasn't it? While we were acting?'

'She could've,' Jackus said, puzzled. It did not make sense, but he knew that he had not been aware of her all the time.

Ann collapsed into a burst of hysterical laughter. 'You don't expect me to believe *that*, do you? It was *you*. You did it while she was talking to me about horrible Mrs Lovett. And you stayed behind to watch me scream when I found out. You're really sick, Colin Jackus. I wish – I wish you'd have a heart attack!'

She was the one who looked ready to have a heart attack. Her breath came in short, awkward gulps and, under his fingers, Jackus could feel her pulse pounding at high speed. Impatiently, he shook her wrists.

'I didn't do it! Why can't you get that into your thick head?'

'Well, who was it?' She wavered for a moment, holding her breath.

33

'Oh, I dunno.' Wearily, Jackus let go of her. He could not think of any sensible explanation. 'Perhaps it was a ghost,' he said flippantly.

For a moment, he thought she was going to hit him after all. Instead, she made an odd sound, something between a scream and a sob, and charged for the door, slamming it after her.

Jackus stared down at the repulsive mess on the floor. He did not really want to touch the remains of the food, but it seemed sensible to clear it up. If he left it, there would be questions and, inevitably, it would end up with more trouble for him. Feeling queasy, he wrapped his handkerchief round his hand and picked up the chewed sandwiches, the gnawed Mars bar and the apple, whose bitten surface was slowly turning brown. He dropped them into the wastepaper basket and then tipped the sandwich box upside-down, banging the bottom of it so that the soggy contents fell into the bin without his touching them. As he banged, he muttered, 'Stupid – hysterical – girls. Fuss – about – nothing.'

But he did not eat his own sandwiches.

Monday, 16th November

'. . . I could feel her there. Like when someone you know stands behind you and whispers, and you know just what they're looking like. I *know* what horrible Mrs Lovett looks like. She's fat and depressing, with that yellowish-grey skin that always seems to have dirt in the creases. Miss Lampeter keeps going on about how fierce and cruel she is, but that's because she doesn't understand. It's not like that. She does horrible things because she doesn't care much about anything or anyone. And because she's frightened of Sweeney Todd. She'd like to escape from him, but she just can't make the effort.

Oh, wouldn't the Lamppost be proud of me! *Such* a good pen portrait of the character I'm playing! Perhaps I'll be a great actress after all!!!

I don't know why I'm pretending it was funny. It wasn't, at all. I was quite scared. Only it doesn't seem so important now, because of the vile, *revolting* thing that happened next.

When I came to eat my sandwiches, someone – *and* no prizes for guessing who – had bitten all of them. I was nearly sick on the floor when I saw it. All chewed and slobby with spit. He must have taken huge bites. Like someone starving.

What made it worse was that he's so dirty. He *smells*. I noticed it before, when we had the workshop and we were sitting round in the dark. A smell of old food and sweat. And it was even stronger today while we were acting. The thought of his picking up my sandwiches and putting them in his disgusting mouth and – YUCK!

I don't think I could have stayed at school for the afternoon if it hadn't been for Marshall. He was really nice. When I ran out of the room, I bumped straight into him, and he asked what was the matter.

I nearly didn't say. I'd really had enough of boys for one day. But it all came flooding out. I couldn't help it. And he was so *kind*. He took me along and gave me a drink of water and made me tell him everything again, slowly. When I got to the bit about Jackus saying it was a ghost, he laughed and laughed. Somehow, it didn't seem quite so bad then.

He said, 'So Jacko thinks it's funny making jokes about ghosts, does he? Perhaps we ought to see he gets some more.' And he gave me a really nice, friendly smile. I never thought he liked me.

Well, when he told me what he'd thought of, I could have kissed him. (I don't mean *really*, of course!). We planned it out straight away and I made a list of jokes in my notebook. And I don't *care* if it makes Jackus get into trouble and lose his part in the play. He'll be the idiot then.

HA, HA! HORRIBLE, CREEPY JACKUS!'

Chapter 4

It was two or three days before Jackus had to attend his
first proper rehearsal. Rather unexpectedly, he found
himself looking forward to it. After the scene with Ann
in the Geography room, he had been waiting for some
kind of trouble, but there was nothing. If anything, she
had gone out of her way to speak to him more than
usual, but she had not referred to the sandwiches. He
assumed that she must be feeling ashamed of the fuss
she had made. Relieved, he pushed the oddness of the
incident to the back of his mind. There was no point in
worrying about it. Instead, he occupied his free time
making sure he knew his lines.

Because he was beginning to see where the interest
of this play business lay. There was something in-
triguing about being able to put yourself inside somebody else's head, to say the words not as you would
have done yourself, but as cheeky, self-confident Jarvis
Williams would have said them. Because Jarvis Williams

was not a bad sort of person. He did not have to worry about what anyone thought of him. He did not have to make the silly jokes Colin Jackus made to stop everyone getting at him. He just bounced on his way, brewing up plans to get the better of Sweeney Todd and assumed that people would join in. And they did. They liked him. Yes, Jackus thought to himself. He was going to enjoy the next rehearsal.

And it started well. They began with a scene where Sweeney was bullying Mrs Lovett, telling her that she could not give up the business of making human pies.

'There will be more – more *meat* soon,' crooned Marshall. 'And you must see to it, Mrs Lovett my dear, that you have another baker ready to deal with it. Otherwise – you know what will happen.'

'Yes. Yes, of course, Mr Todd,' stammered Ann, the very picture of a fat old woman sweating with fright. In the wings, Jackus chuckled to himself. Ann might not like Mrs Lovett, but there was no doubt that she could do her very well now.

The next scene was his first. He waited, listening to Ann selling pies, ready for the words that would tell him to enter. But just as Ann started to speak them, he found Stephen at his elbow.

'I say, old bean,' Stephen muttered, 'what do short-sighted ghosts wear?'

In spite of himself, Jackus was distracted. He turned to look at Stephen, who gave him an owlish grin and whispered, 'Spooktacles!'

At the same moment, Miss Lampeter's voice, slightly irritated, rang out from the front row of seats. 'Colin! It's your cue. You're late.'

With a scowl at Stephen, Jackus hurried out on to the stage. But he was rattled. Somehow, he had lost the feeling of Jarvis Williams, and it took him a few minutes, while he fumbled his first couple of lines, to remember the right voice, the ingratiating grin. Just as he recovered himself, Ann said, in her gruff Mrs Lovett voice, 'But have you no friends or relations to take care of you?'

Before he could reply, she turned slightly away from the audience and murmured, 'And what did the big ghost say to the little ghost?'

Taken completely by surprise, Jackus stuttered and stopped. With a triumphant grin, Ann breathed, 'Don't spook until you're spooken to!' and turned back to the front again, the picture of innocence.

'Colin!' shouted Miss Lampeter, exasperated. 'Will you please concentrate!'

'Ann put me off,' he said sulkily.

'Nonsense!' Miss Lampeter frowned. 'She was acting perfectly well. Don't make excuses. Now go on, and think about what you're doing.'

But it was very hard to think about what he was doing. Whichever way he turned, faces peered out of the wings at him. Voices, pitched just high enough for him to hear, told silly jokes about ghosts. He knew that something was going on, of course. Someone had organized this, because of what he had said about the sandwiches. But it was almost impossible to ignore. The faces always appeared when he was least expecting them, making him stumble idiotically. And each time, he heard Miss Lampeter's quick, annoyed sigh.

Finally, Ann left him alone, supposedly to look round the cellar and, gritting his teeth, he determined to get through the rest of the scene without a mistake. And he nearly did it. But just as he came to his big moment, when he stole one of the pies, bit into it and found a button, he heard a hiss from somewhere near his feet.

'What do ghosts have for lunch?'

Unable to stop himself, he glanced round. There was Benny, lying on the floor in the wings, shielded from the front of the stage. He smiled up as Jackus faltered, the innocent smile of a baby, and whispered, 'Spook-ghetti!'

With an enraged snort, Jackus pulled a face at him. Simultaneously, he heard Miss Lampeter explode.

39

'Colin Jackus! I don't know what you're playing at, but you seem to be doing your best to ruin this rehearsal. You've hardly said a single line without making a mistake. And I particularly asked you all to learn your words for today.'

'It's not my fault!' Jackus protested. 'It was –'

'*Don't* make excuses!' she snapped. 'I'm not stupid. I know when someone's not trying. We'll have to skip the rest of your scene, and I'll hear your words tomorrow, at lunch-time. I'll write it down to make sure I don't forget.'

Jackus began to walk off the stage resentfully, avoiding the grinning faces all around him. As he reached the wings, he heard Miss Lampeter give another sigh. In a different voice, she said, 'Oh, what a nuisance!'

'What's the matter, miss?' Ann sounded as virtuous and helpful as usual. No one would think, Jackus reflected sourly, that she had just helped to ruin the last two scenes.

Miss Lampeter gave a slightly embarrassed laugh. 'It's silly, but I seem to have lost my silver propelling pencil. I was using it to make notes with, and now I can't find it anywhere. Has anyone picked it up or knocked it on the floor?'

Obligingly, Ann jumped down from the stage and began to grovel around looking under the chairs, her fat bottom sticking up into the air.

'I can't see it anywhere.'

'Oh dear.' Miss Lampeter frowned. 'Look, I don't want to make a fuss, but it was my father's, and I'm fond of it. Do you think it could have got into someone's bag or someone's pocket? By mistake, of course. Can you all look?'

Most people began to turn out their pockets and rummage through their bags, but Jackus did not move. He knew that he had not touched any pencil.

'Benny?' Miss Lampeter said. 'Mandy? Stephen?'

She glanced round at the members of the cast in turn

and, one by one, they shook their heads or shrugged. All at once, Jackus realized that she was purposely avoiding his eyes. Tactfully not accusing him. Only it was worse than an accusation. It was so obvious to him that he imagined everyone else must see it too. Furiously, he dragged his bag out of the wings and on to the middle of the stage. With a defiant gesture, he tipped the contents out on to the floor and then turned his pockets inside out, scattering pennies and conkers all over the floor.

'I haven't got it!' he said loudly. 'Look. You can frisk me if you like.'

Rather to his surprise, Miss Lampeter went pink. 'Oh, Colin. I didn't mean –' She stopped. 'It's all right, dear. Put your things away.'

Jackus stood by his bag and glared at everybody. Then, scowling, he bent forward to pick up his books.

At that moment, without warning, something silver whizzed past his right shoulder and down into the body of the Hall, narrowly missing Miss Lampeter's face. It fell to the ground behind her with a clatter. Jackus whipped round to see where it had come from. But there was nobody behind him.

In total silence, Miss Lampeter put a hand up to her cheek, and Jackus saw the fingers tremble. Then, pressing her lips together, she bent down to pick up the pencil.

'That was very dangerous, Colin,' she said, in a voice that shook with anger. 'It could have hit me in the eye.'

'But I didn't –' Jackus said. And stopped. He could tell, from the scorn on the faces in front of him, that no one was going to believe a word he said.

But Ann spoke reluctantly, forcing the words out. 'I don't think it could have been Jackus. I was watching him. He couldn't have –' She swallowed hard and stared at the ground.

''Course it was Jackus,' Benny said loudly. 'Who else could it have been?'

41

Miss Lampeter looked at him, still furious. 'Well, Colin?'

Suddenly, Jackus was sick of the lot of them. He had not done a thing, and they had all been against him, all the time. Messing up the rehearsal when he had been looking forward to it.

'Must have been another ghost, mustn't it?' he said sourly.

Miss Lampeter banged her hand down impatiently on the chair beside her. 'That's *it*! I've had about as much of you as I can take for one day. First you mess up my rehearsal, then you steal my pencil, and then you attack me with it. It's all completely childish and pointless, and you haven't even got the grace to apologize. You can just go down to the library and sit and learn your lines while the rest of us get on. And I'll come down at the end of the rehearsal and make sure you're still there. Now go away.'

There did not seem to be any use in arguing. Angrily, Jackus stuffed the rest of the things into his bag and stamped off the stage. As he came into the wings, he passed Marshall, who was lounging with his hands in his pockets, smiling.

'Load of stupid kids they all are,' Jackus muttered. 'Can't think how you stick them.'

Marshall clapped him on the back. 'It's tough when you don't fit in,' he said lightly.

Jackus pulled a face at him and strode off down the Hall. But he felt slightly less annoyed. At least there was one person in the cast who had enough sense not to play stupid games. At least Marshall was sympathetic.

Chapter 5

When Jackus pushed the door open, the library was empty, and shadowy in the half-darkness. He flicked a switch and sudden light gleamed off the polythene spines of the books on the low shelves and showed up the tattered red covers of the encyclopaedias in the tall bookcase at the end of the room. Through the big windows, he could see the last few people straggling home down the drive. Mrs Russell, the needlework teacher, was harrying them and she looked curiously at Jackus as he stood alone in the middle of the library.

He could not face any more trouble. It would be wise to look busy. Sitting down at one of the tables, he opened his script, trying to appear studious. There was no point in going over his own lines. He knew them backwards. Instead, he flicked idly through the pages, reading some of the scenes in which he had no part. His eye fell on a speech of Marshall's. A satisfyingly cruel, creepy speech, arranging for his apprentice – who was Benny – to be shut up in a madhouse. The idea pleased Jackus. Remembering that little fair face grinning up at him as Benny got him into trouble, he thought that there was nothing he would have liked better than to see it safely behind bars.

He shut his eyes for a moment, conjuring up a picture of Jonas Fogg, the madhouse keeper. It should have been a picture of Stephen, by rights, but Stephen's face was too mild. Jackus imagined a beaky, hawklike nose and piercing black eyes. Then he added a thin moustache and a whip in the right hand. Serve Benny right.

'I have brought you my apprentice, Mr Fogg,' he snarled aloud. 'His name is Tobias Ragg, and I beg you to keep him in close confinement. *Very* close confinement.'

The words sounded satisfyingly sinister as they echoed through the empty room. Jackus grinned. He could see why Marshall liked playing Sweeney Todd. It must be nice to have that kind of power over your enemies. Working more cruelty into his voice, he went on with relish.

'Kind treatment will be no use. He must be chained up and fed on bread and water. That's the only way to keep these stubborn apprentices in order.'

He paused to draw breath and, in that second, he heard a faint noise. Like a suppressed, wordless whisper.

'Hallo?' What an idiot he must have sounded if there were someone hiding in the room. Getting up, he peered round behind the bookcases, but there was no one there. Must have come from outside. Staying on his feet, he began to stride round the room, starting on Marshall's next speech.

'I will pay you to look after him for six months, Mr Fogg. But I do not suppose – that he will last so long. He –'

44

This time the whisper sounded from behind him, clearer but still wordless. He spun round, looking straight at the corner it had come from.

There was nothing there.

'Who is it?' he said softly.

Instantly, the whispering stopped short, with a quick gasp of breath, as if the whisperers were terrified at being overheard.

Jackus gulped slightly and tried to be sensible. *Tape recorders*, he thought. There must be little cassette recorders hidden behind the books. He sidled towards the shelves near where the last whispers had sounded and pulled out a few books, dropping them on to a table. But all he could see was the bare wood at the back of the shelves. And, behind him, the whispering had started up again.

He rubbed the damp palms of his hands against his trouser-legs, suddenly furious. It must be some kind of trick. It *must*. Why couldn't they just leave him alone? He yelled angrily, 'Shut up! Shut up at once!' Scrabbling round his mind, he tried to think of something frightening to say. 'You all deserve to be shut up in a madhouse. Chained and beaten and fed on bread and water!'

Abruptly, all sound stopped, chopped off. He could hear nothing but the tingling of silence in his ears. But all over his skin, he could feel the cold, ominous pressure of empty air.

Then, slowly, something beside him moved. He looked sideways. One of the books he had pulled out lay open on the low table. One by one, the pages were turning over, not in flutters, as the wind would blow them, but deliberately, as if fingers separated and tweaked them.

He reached out a trembling hand and snapped the book shut, viciously.

Immediately, something whistled past his ear and thudded on to the ground in front of him. Stupidly, he stared down at the cover of the book which lay on the floor. *Birds of Britain and Europe* said the embossed letters, but his mind did not seem able to take in the sense of them. Very slowly, he turned round and, as he did so, he saw the books begin to slide from the shelves. Gently at first, and then more and more violently, all over the library, they slithered down, their hard covers hitting the wooden boards with a steady smack, smack, smack.

Jackus started to pick them up and jam them back on to the shelves anyhow, his fingers shaking so much that they could not find the right spaces. And, without meaning to, he was yelling.

'Stop it! Stop it, whoever you are!'

Then from the tall bookcase at the end of the room, his eyes caught a movement. In a red arc, the whole line of encyclopaedias was falling, neatly, one volume after the other. The covers were bright in the air, and the pages exploded in a white, fluttering cascade as the books hit the ground and spilled open. He darted forwards to try and catch them and, suddenly, met a resistance. Nothing he could have described. Just something which made it impossible for him to step forwards.

And then the big bookcase started to fall. Quite slowly, it tipped forwards, sending the rest of the books on its shelves thundering down chaotically, first from the top and then from the bottom, until the great wooden stack was empty. It pitched on to its front, hitting the ground with a sonorous rumble.

In the total silence which followed, Jackus heard feet outside the door. He heard the door open and someone come in. But he no longer had the power to turn his head. He was certain that he would see nothing. Nothing, or something worse than nothing.

The voice which spoke was reassuringly familiar and icy.

'Well, Jackus?'

'Sir.' He turned shakily and stepped towards Old

Garner, unbelievably pleased to see the headmaster's sour, lean face.

'What have you been doing, boy?'

'I didn't,' Jackus said weakly. He felt too peculiar to protest properly.

'You *didn't*?' Mr Garner's grey eyebrows rose. 'What happened, then? Did a crowd of hooligans wreck the place and then jump out of the windows?' His eyes flicked over the closed catches. 'Or did the books simply tumble from the shelves unaided?'

'Yes,' Jackus said softly. 'Yes, they did.'

He did not expect to be believed. Miss Lampeter might have risen to that one. She would probably have sat down, her eyes bright with the desire to understand, and made him tell her what had happened. Old Garner merely stared at him for a moment and then glanced round at the tumbled shelves and the books which littered the floor.

'Well,' he said at last, 'they may have fallen by themselves, but they won't grow legs and climb back, will they? We'd better see if we can lift the bookcase.'

Numbly, Jackus followed him to the end of the room and the two of them bent down to lever up the tall stack of shelves. It was heavy and awkward and they strained and struggled before it finally moved up, hitting the wall with a shudder that knocked a hole in the plaster and sent a cloud of white dust down on to Mr Garner's bald head.

Panting heavily, he took out a handkerchief and mopped the dust away, then took off his glasses and cleaned them carefully. His face was white, and Jackus could hear him catch his breath with an effort before he spoke.

'Now,' he said, as soon as he had recovered sufficiently to speak in his usual controlled voice, 'there are the books to be dealt with.'

'Yes, sir,' Jackus murmured, feeling oddly concerned. Part of him wondered what would happen if Old

Garner suddenly had a heart attack, and dropped down dead. Would he be charged with manslaughter?

But Mr Garner did not allow himself to look feeble. He jammed the glasses back on to his nose and flicked a long, sparse strand of hair into place across his scalp. When he spoke, it was with ferocious briskness.

'You will put all the books back on the shelves, Jackus. Carefully and in order. Checking the numbers on the spines. And then you will wait for me to come and check that you have done it correctly before you go home. Is that clear?'

'Yes sir,' Jackus muttered.

'That's all,' Old Garner said. 'I do not propose to punish you further. I think that, by the time you have set the library to rights, you will think twice before causing such chaos again. Now get to work. I shall be back in half an hour.'

He stalked out of the library and Jackus looked round in despair at the piles of books littered everywhere. A small corner of his mind was full of admiration. If he had really done it, if he had wrecked the place, in a fit of rage, it would have been the ideal punishment. But he had not done it.

Groaning to himself, he picked up a book, read its number and slid it into place. As he did so, there was a faint stirring in the air by his ears, gentle and unobtrusive.

'It's your fault,' he said bitterly, not really feeling that he was speaking to anyone. 'You landed me in this. Why can't *you* pick up the wretched books and put them back?'

For a ridiculous moment, he thought it might work. That all the books might fly up magically and jump back into their places, as though nothing had happened.

It was not like that, of course. The books stayed obstinately still. Instead, from different corners of the library, he heard quiet, weary noises, little breaths, as though a hundred children yawned in inexpressible weariness, after a hard day's work. It was eerie, but not threatening and after what he had been through, he was actually able to smile at it.

'Go on, then. Go to sleep. I'll do your dirty work for you.'

Feeling suddenly relaxed and peaceful, he picked up an armful of books and began to sort them. The next moment, he knew he was really alone. There were no more noises. Nothing else to do – except for the long slog of reading titles, finding numbers, clearing the floor.

Half an hour was not really enough time, but he worked fast, and all the time he worked he was thinking, puzzling over what had happened. Every time he tried to work out some sensible explanation for what he had seen, his mind jumped away, refusing to work.

There had been no one else in the library. He was sure of that. And there were no wires, no springs, no cunning bits of mechanism. The empty shelves were empty and the books had nothing hidden in them.

And there was something else. The resistance that had stopped him going forward to catch the encyclopaedias as they fell. That had not been imagination. Something outside himself had stopped him. And if it had not, he would have been lying underneath the heavy bookcase when Old Garner came in. Certainly injured, and possibly dead. But he had been saved. There had been anger at work in the library, from the very moment he had started to speak Sweeney Todd's words. But it had not been directed at him. Not himself.

Bending and stretching, climbing the library steps and arranging the books, Jackus brooded, gnawing his bottom lip and humming to himself.

When Old Garner came back, he was standing in the middle of the floor, just finished. The headmaster glanced round and nodded.

'Right. You seem to have worked hard. What's that book you've got in your hand? Can't you find the place for it?'

'It's not that.' Jackus looked down at it. 'I found it when I was sorting. Can I take it out, please?'

'Hmm. Don't see why not.' Mr Garner tossed a library slip across to him and, while he was filling it out, picked up the book and looked at it. *The Complete Ghost-Hunter's Guide*. Are you being impertinent, boy?'

'No, sir,' Jackus said evenly. Then, in spite of Old Garner's frown, he risked a question. 'Do you believe in ghosts, sir?'

For a moment the old man said nothing, weighing the book in his hands. Then he looked up. 'I believe in evidence. In not making up my mind before I have gathered sufficient information. An old-fashioned belief, perhaps, but it works well.' His eyes were shrewd as he scanned Jackus's face. 'And I'm always ready to listen to evidence. Honestly given.'

It was half a question, but before Jackus could say anything, there was a brisk tapping of feet along the corridor. The library door was pushed open and Miss Lampeter began to speak even before she was properly in the room.

'Now, Colin, I hope you've done what I told you, and – oh.' She broke off as she saw Old Garner.

'Good afternoon, Miss Lampeter,' he said gravely. 'You were the person who sent Colin down here, were you? I was a little surprised to see him. I had thought he would be at your rehearsal. I trust he has not been mis-behaving himself.'

Here we go, thought Jackus. *More trouble*. But, to his surprise, Miss Lampeter went a little pink and shook her head.

'No, that's quite all right, Mr Garner. I can deal with Colin,' she said brightly. 'He just came down here to learn his lines. Didn't you, Colin?'

He was so amused that he nodded without thinking.

Fancy her not wanting to tell Old Garner what a shambles the rehearsal had been.

The headmaster himself looked as wooden as ever. 'Well,' he said gravely, 'you can rest assured that Jackus has been working hard while he has been down here. Very hard indeed.' There was not even a slight smile. Nothing to give the game away. At least you could trust Old Garner, Jackus thought grudgingly. When he said picking up the books would be the end of the matter, he obviously meant it.

'I'm pleased to hear that,' Miss Lampeter said enthusiastically. 'I'll hear your words tomorrow, Colin. Before lunch. After school we're all going to the Art Room to start painting the scenery, so don't forget to warn your mother.'

Ignoring the headmaster's slightly startled expression, she turned and clipped out of the room on her spiky heels. Jackus scooped up *The Complete Ghost-Hunter's Guide* and moved towards the door. But he was stopped by the headmaster's voice.

'Jackus.'

'Sir?' Glancing over his shoulder, he saw that Old Garner was studying him with a thoughtful expression.

'Was there something that you were going to say to me? Just before Miss Lampeter came in?'

But Jackus had had enough. He just wanted to go home and forget about school and the play for as long as he could.

'No sir,' he said blankly. 'I don't think so.'

There was a pause, and then Mr Garner smiled faintly. 'I beg your pardon. I must have been mistaken. Goodbye then.'

'Goodbye, sir.' And Jackus hurried out of the building, feeling as though he had come out of somewhere close and stifling into the blessedly ordinary fresh air.

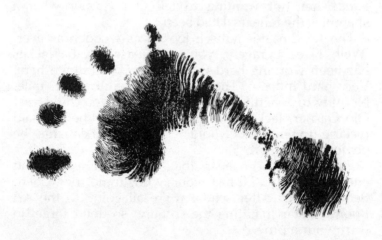

Chapter 6

'*Everyone* must hate Marshall,' Miss Lampeter said decidedly, slopping blue paint on to the backcloth.

Miss Stevenson, the art teacher, sniffed and leaned over sideways to fill the gaps she had left. Without noticing, Miss Lampeter went on gaily, flourishing the paint brush and flicking paint all down her apron.

'I want you all to feel that you're in league, held together by fear and hatred. It must come across to the audience like a great black wave. Every time they see Marshall, they'll hiss, but they must be shuddering as well. Isn't that right, Marshall?'

She turned round, still on her knees, and smiled at him. He was the only one of them not working. Hands in his blazer pockets, he lounged back on a desk.

'Just keep on painting, all of you,' he growled. 'I don't want to see anyone slacking.'

'Oh, worse than that!' Miss Lampeter murmured

cheerfully. 'You've got them all in your power. Can't you manage to be a bit more beastly?'

Marshall slid elegantly off the desk and started to walk round the art room, peering over the shoulders of the painters.

'You! Tobias Ragg!' he snarled at Benny. 'If you don't do what I say, you'll land up in a madhouse. And I'll have your mother arrested for stealing.'

'Oh no, Mr Todd, please!' squealed Benny, enjoying himself. 'Not Jonas Fogg's madhouse!'

'The very same,' Marshall sneered. 'I have him in my power too. Haven't I, Jonas Fogg?' He whirled round and caught hold of Stephen's ear. 'I know what became of the other apprentices I gave into your care. Put a foot wrong, and I'll have the law on you.'

'Ouch! Yaroo!' Stephen said absently. He was absorbed in painting a lamp standard and was not really paying much attention to Marshall.

'Look at me when I'm speaking to you!' Marshall rapped out. He gave the ear a vicious twist and Stephen's arm jerked, sending a fat, black squiggle down the canvas.

'Oh, shove off, Marshall!' He was annoyed. 'I was concentrating.'

'Yes, do be careful,' Miss Stevenson said fretfully. 'We can't afford to buy any more of the cloth.' She frowned and Jackus, working away at the bottom of her section, heard her mutter, 'We've never had the *cast* painting the scenery before.'

Miss Lampeter picked it up, too. She grinned unrepentantly. 'But don't you see? What I want is total involvement. I want them to eat and dream and breathe Sweeney Todd. When they wake up in the middle of the night, I want them to think they're in the nineteenth century.'

'I'm sure we'll all have nightmares about Marshall,' Jackus said politely. He was painting bricks, alternating shades of yellow and grey, and with his brush he sketched a devil, complete with horns and a tail, before he filled the next brick in.

'You be quiet, Jarvis Williams,' Marshall hissed in his ear. 'A puny little apprentice like you can't interfere with *me* – and you've made a right mess of that brick, haven't you?'

His voice finished up unexpectedly sharp and Jackus glanced up at him. 'Trying to needle me properly, are you?' He gave Marshall a smile of sugary sweetness, to show that he was not going to be tricked into playing that game. Let Miss Lampeter set the others against Marshall if she wanted to. *He* was not going to be conned. And Marshall grinned back. That was the good thing about him. Even if he did not always co-operate, he knew what was going on.

He did not waste any more time talking to Jackus. Instead, he moved round the other groups, disturbing them and finding fault with their painting. It was not quite the joke it had started as. All over the room, people were growing angry and resentful. The whispers were annoyed. But Marshall went on sauntering about, with an expression of evil relish on his face.

As he reached Ann, he leaned casually over her shoulder and said, in his normal voice, 'That's not really very good, is it?'

'What do you mean?' She looked up, frowning earnestly. She had been working hard while the others grew restless, and there was a smudge of black paint on the end of her nose. 'I've done five window frames, and they're very tricky with those little criss-cross bits.'

'And you've gone over the edges on every one,' Marshall said gravely. 'Still, I suppose someone as fat as you is bound to be clumsy.'

For a moment, Ann was clearly speechless. Then she said, 'Oh, don't be a *beast*, Marshall.'

'Don't be a *beast*, Marshall.' He mimicked her voice. 'I can see you don't like the truth, Mrs Lovett, my dear.' He patted her annoyingly on the head, and her brush spattered paint down the front of her blouse.

'Stop it!' Her voice had risen, upset, and the art teacher moved automatically, about to tell Marshall not to make a nuisance of himself. But Jackus saw Miss Lampeter put a finger on her arm and shake her head. She was watching Ann and Marshall intently.

'Now look what you've done,' said Marshall reprovingly. 'I told you you were clumsy.'

'Oh, leave me alone.' Ann's bottom lip trembled, as if she were about to burst into tears. Still Miss Lampeter did not move, and still she held Miss Stevenson back.

Inevitably, the others began to join in.

'Keep off her back, Marshall.'

'Yeah, quit annoying her. She's working, and you're not.'

'Dear, dear,' crooned Marshall irritatingly. 'You're all making very heavy weather of a simple little job, aren't you?'

He jerked the paintbrush out of Ann's hand and, with maddening ease, drew in four or five of the little lines over which she had been toiling. 'You'd all be finished by now if you weren't as clumsy as Little Orphant Annie here.'

Ann burst into tears and Stephen and one or two of the other boys leaped up to drag Marshall away from her, shouting at him. Jackus sat back on his heels and watched in amazement. The room was in an uproar, for no real reason. People were jumping up, yelling, even punching each other. And there was the Lamppost grinning idiotically as she watched it all.

The art teacher jerked her arm free and got to her feet, but before she could do any more one of the paint pots went flying across the room, leaving a trail of yellow paint on the floor.

'Now look what you've done!' Miss Stevenson was adding to the yelling now. But no one took any notice of her. Instead, a tin of red paint, which was standing on a desk, tipped over on its side and began to pour out a steady, scarlet stream.

Jackus felt the back of his neck prickle. Everyone else was fighting or shouting or blaming someone. Only he was watching coolly, and he was almost sure of what he had seen. The next moment, he *was* sure. The black paint, alone by itself in the middle of the floor, tipped gently over, untouched by anyone. A dark pool of stickiness gushed out and began to spread wide. Then, as if it had been knocked again, the tin rolled across the floor, clattering. Jackus knelt and stared at the spreading paint.

Miss Stevenson was just beginning to get control of the struggling mass of people. She pushed them into chairs, one by one, bellowing at the top of her voice.

'Sit down, all of you! I've never seen such behaviour! I won't have any of you in the art room for the rest of the term.'

Panting, they lapsed into silence and sat glaring at Marshall who, somehow, still seemed unruffled. Casually, he looked round at the mess, clicking his tongue against his teeth. One or two of the boys stirred in their chairs and Miss Stevenson turned purple.

'Don't you dare move!'

Scowling, they sat straining forwards at Marshall, as if they were held by invisible ropes.

'Oh, that's perfect!' breathed Miss Lampeter into the silence. 'Just the mood I want. You hate him, but you daren't touch him. If you can get that into the play, it will be brilliant.' Then she laughed. 'But – oh dear, I'm afraid Miss Stevenson's right. We've made a terrible mess. Someone's going to have to stay behind and clear it up.'

'I'll do it,' Jackus said quickly. 'I don't mind.' He could see everyone staring at him in amazement, but he did not explain. And he kept his eyes well away from the strange thing he had noticed in the corner of the room.

'Oh, Colin, you don't have to.' Miss Lampeter sounded surprised as well. But pleased. 'I think you were the only person not joining in.'

56

'I don't mind,' Jackus said again. 'Just get everyone out of here, so I can start.' To his amusement, he noticed that even Marshall was looking puzzled.

'I'll help you,' Ann said. Inevitably. 'It was my fault really, I suppose.'

'Don't bother,' muttered Jackus. But he knew there would be no way to put her off. She sat there like a lump, while Miss Lampeter shooed the others out of the room.

'Mind you don't tread in the paint, now. You'd better take off your shoes when you get to the door. Just to be on the safe side. You can wipe them on the grass when you get outside.'

One by one, people crept towards the door, cowed by the sight of the chaos. There was a traffic jam at the door, as everyone hopped about, taking off shoes and plimsolls, and then the sound of stockinged feet padding away down the corridor, without any of the usual chatter.

For the first time, Miss Lampeter looked embarrassed, as she glanced at her watch. 'Look, I've just realized. I've got to meet someone at half-past five. Of course, I'll stay and help clear up if you like, but –'

'Never mind,' Miss Stevenson said wearily. 'You go off. These two can stay until it's done.' She gazed round in despair at the rainbows of trampled paint. 'Take at least an hour, I should think.'

'Tell you what, miss.' Jackus tried to sound bracingly sensible. An Ann-voice. 'You go off to the staff-room and have a cup of tea. That'll make you feel better. We can get started without you.'

'Why, thank you.' Miss Stevenson looked surprised but grateful as she passed a hand across her forehead. 'I think I will. Shan't be long.'

As she walked off down the corridor with Miss Lampeter, Jackus thought triumphantly, *Only one to go*.

'You go and find some buckets and water and stuff,'

he said quickly to Ann, before she could start to organize him. 'I'll pick up the tins.'

She wavered for a moment, while he longed to give her a hearty shove. Then she nodded. 'Water'll get it off, will it?'

'Should do. Oh, go *on*. Find as many buckets as you can.'

That should keep her busy for a bit. As she hurried out, Jackus began to tiptoe carefully across the room, trying to keep his feet out of the paint. It was over here he'd seen it. By the radiator.

At first he thought he had made a gigantic boob. Saddled himself with all the clearing up for no good reason. There were certainly plenty of footprints on the floor, but they were all ordinary ones, showing the patterns on the bottom of shoes, or the unmistakable diamonds of plimsoll soles.

Then he saw it. Just as he had caught sight of it out of the corner of his eye while the others were being stopped from fighting. The clear print of a small bare foot, with the line of toes showing plainly. As he gazed down at it, he felt a slow shiver wobble its way across his shoulder-blades.

And there was another, over by the empty tin of black paint. But a different one this time. Still small, but with the little toe damaged, so that the foot ended in a kind of jagged lump.

'People with bare feet,' he murmured aloud. 'Children. Knocking the paint over and tracking round in it. But *why*?'

His voice sounded flat in the empty room, echoing back at him from the wooden easels and the glass of the uncurtained windows. Outside, it was almost dark, although it was early, and the windows showed nothing but reflections of the mess inside.

Slowly, Jackus ran a finger round one of the prints. Then he had an idea. With a quick, careful glance at the door, he picked his way through the paint to the shelves where Miss Stevenson kept the paper, stacked in neat pastel piles, pink, pale blue, pale green, white and black. Sliding out a white sheet, he carried it over to the first footprint and laid it carefully across the top, smoothing it with his other hand. When he peeled it away, there was a perfectly-defined copy of the footprint. Swivelling round, he took another copy, of the other footprint with the toe missing. Then he took out his handkerchief and scrubbed at the floor with it, so that the toe-marks disappeared in an untidy smear of paint. Now the only proof that they had been there was the prints on the paper, plain and perplexing.

From outside the door came a metallic clank. Quickly, Jackus slid his piece of paper behind a radiator. Then he began to scoop up the scattered paint tins. By the time Ann staggered through the door, slopping water everywhere, his arms were full.

'You kept the easy job for yourself, didn't you?' she said. 'It took me ages to find these two buckets, and my arms are dropping off with the weight. Here, take one.'

She lowered them carefully to the floor and tossed him a cloth. Wringing it out in the water, he smeared it in a great arc across the paint-covered floor.

'Not like that,' Ann said scornfully. 'You'll just make a worse mess. Look, like this.' Carefully and efficiently, she began to scrub away at the edge of a patch of blue. 'It comes off all right, anyway.'

'Told you, didn't I?' More gently now, Jackus began to work on his own side of the room, whistling quietly between his teeth and not looking at Ann. He did not want her chattering away in his ear. He had things to think about.

Because he was not watching her, he did not notice at first when she sat back on her heels and stared at the floor. Picked up her cloth. Put it down again. Went on staring.

'Jackus,' she said suddenly, 'come and look at this.'

'What now?' he glanced round at her. She was on the far side of a table, looking down.

'I've found something rather odd. Come and see.'

'What is it?' As if he had not guessed already.

'It's –' she laughed, nervously, '– it's a footprint.'

'Honestly, you are thick. Of course it's a footprint. There were enough people crashing about to make five million footprints.'

'No, look. It's a *bare* footprint. Why won't you come over here?'

He lounged round the desk and looked down. It was the one with one toe missing. Definitely peculiar. ''Course it's a bare footprint,' he said briskly. 'The Lamppost told them to take their shoes off, didn't she? Why are you making such a fuss?'

'But they didn't take their shoes off until they got to the door.' Her eyes were big and round as she gazed up at him. 'It's really odd.'

For a moment, he was tempted to tell her that it was not the only odd thing that had been going on. She would have believed him. He could tell that, just from looking at her face. But then, in time, he remembered Marshall. What would Marshall say, if he knew Jackus had been confiding in *Ann Ridley*? He'd laugh his head off.

'Don't be soft.' Leaning over, he swished his cloth across the floor, rubbing the footprint into an indistinguishable blur. 'We've got enough to do, without you playing games. Get on with it.'

Turning his back on her, he started to scrub away at another patch on the floor.

Monday, 30th November

'. . . I thought it was a game, at first. Marshall was calling people by their names in the play and they were pretending to be scared. But then it all seemed to get nastier. Mandy was really annoyed when he told every-one she had a spot on the back of her neck. I mean, it wasn't as though we hadn't all seen it, but you don't *say* things like that. But I still thought he meant to be joking.

But what he said to me was really horrible. Unkind. He said I was fat. And when I looked at him, I could see he'd said it because he knew it was the thing that would annoy me most. (I *knew* Mum was wrong when she said I only look fat to myself. I'm disgusting. I wish I were as thin as Mandy.)

I could hardly believe it. How could he be so beastly when he was so nice to me the other day? And he just grinned and grinned. I could have killed him. I *hated* him. And I went on snapping back at him, as though I couldn't help it. Then all the others joined in, shouting and punching. All the time, I was wishing and wishing Miss Lampeter would stop it.

I know why she didn't though. It's her precious play. That's all she can think about these days. She was so pleased we were hating Marshall that she didn't care about the noise or the mess or anything.

But it's ridiculous. I don't hate Marshall. Of *course* I don't.

Then, when it was all over, Colin Jackus said he'd stay and clear up. Which was really weird. I've never known him offer to do anything before. I said I'd help him, of course. I mean, someone had to, even if it *was* him.

But I wished I hadn't stayed afterwards, because I saw

61

something *really* odd. Something I just couldn't under-stand. *There were bare footprints in the paint.* Jackus said it was because all the others took their shoes off, but I knew that wasn't true. He's just too thick to see it. There was something really peculiar about those footprints.

And that makes three peculiar things. My sandwiches (unless that was Jackus – I'm not quite so sure, now), Miss Lampeter's pen flying through the air and now this. I don't know what's happening to this play, but whatever it is, it's weird, and it's making us all very strange.

I just wish I weren't the only one who noticed these things. If only there were someone I could talk to about them, it might not be so scary. But who could possibly understand?

Well, I know Marshall would. And it would be super if he was nice, like he was about the sandwiches. But if he was horrid, like today, I think I'd die. . . .

Sometimes, I wish I weren't in the play at all.'

Chapter 7

'Another rehearsal tonight, Col?' His mother put his bacon and egg down in front of him. 'It's taking up a lot of time, isn't it?'

Jackus grunted and began to eat his breakfast, wishing she would leave him alone. But she had obviously decided that it would be a good time for a talk. Pouring herself a cup of tea, she sat down at the table.

'And are you enjoying it? Getting on all right with the others?'

'Suppose so.' There was no point in explaining how everyone glared at him when he went into rehearsals, as if he were an intruder. She would never understand. And it did not even seem very important to him, now. Not now he had the piece of paper with the footprints folded up in his pocket. That gave him more than enough to think about. He chewed another mouthful of bacon.

'How's Big Colin?' his mother persisted.

'Oh, Mum. Why do you keep going on about him? He's O.K., of course. He's always O.K.'

'But you don't seem as friendly as you used to be. Rose and I were talking about it the other day. I can remember a time when you were always in and out of each other's houses. But I haven't seen him for months. Have you quarrelled?'

'Of course we haven't quarrelled,' Jackus said shortly. 'Don't be thick. Why should he come round here, anyway? What do you expect us to do? Play trains on the floor?'

Mrs Jackus looked hurt. 'There's no need to shout at me. I know you're not babies any more. But you could talk. What do you think Rose and I do?'

Jackus had had enough. He finished his breakfast and stood up. '*Me*? Talk to Marshall?' he said bitterly. 'When I'm so stupid and he's so brilliant? You know that's what you think.'

'Oh, Col!' She looked distressed. 'I didn't mean –'

But he had already gone, banging the door after him. As he walked to school, he smiled to himself, rather sourly, thinking that it was ironic she should have chosen that morning to raise the subject. Because he *had* been intending to talk to Marshall. He needed to discuss with somebody all the peculiar things that had been happening, and who else would understand? The rest of them would think he was just moaning about the play and trying to disrupt it. Especially after that silly business of the ghost jokes that had got him banished to the library. But Marshall had not been telling ghost jokes. He had only lounged in the wings watching, with a superior smile. And he was clever enough to realize that some very strange things had been going on. Yes, perhaps it would be a good idea to talk to Marshall.

But it was not as easy as that. All day, Jackus watched for an opportunity, and no opportunity came. Whenever they sat down for a lesson, the chair next to Marshall was inexplicably occupied and when break-time came round Marshall had disappeared. At last, Jackus decided that he would have to make an opportunity. Sitting in the French class at the end of the afternoon, he kept glancing across at Marshall, trying to catch his eye.

But Marshall was busy with his own thoughts. Overtly, he was paying attention to the lesson. Whenever a question was asked, he was ready with the answer and when they were set to do an exercise he wrote busily, finishing before Jackus was half-way through. But all the time, his left hand was in his blazer pocket, as if he were turning something over and over. When he finished writing, he took it out and laid it on the desk in front of him, staring down at it with a strange, amused expression.

It was not until the bell sounded that Jackus was able to leave his seat. Immediately he jumped up and walked across the room.

'Marshall?'

'Yes?' Marshall's tone was casual, but his hand moved like a flash to cover the object on the desk, arousing Jackus's curiosity.

'Here, what have you got there?'

'Nothing.' Then Marshall grinned. 'No, why shouldn't you see? Everyone will get a look at it at the rehearsal.' He moved his hand away, revealing a short length of polished wood with a metal rivet showing shiny at one end. It was like a handle, rubbed dark and smooth with long use, and Jackus stared at it, puzzled.

'Never seen one of these before?' Lazily, Marshall flicked at the edge of it and a long, gleaming blade unfolded, its cutting surface ground to a hollow sharpness.

'What's it for?' Jackus reached out a hand and touched it. The metal was cold under his fingers.

'It's for the play, of course.' With a quick movement, Marshall scooped it up and got to his feet. 'You'll see. At the rehearsal. The Lamppost'll be over the moon.'

And he was gone, leaving Jackus baffled. It was only when he started to follow down the corridor that he remembered he had meant to talk to Marshall about something completely different.

In the Hall, Miss Lampeter was bustling about on the stage, setting out furniture. She was working with more than her usual speed and her hair was tumbled about, hanging in untidy strands over her collar. As she worked, she was calling over her shoulder to the people who came into the Hall.

'We'll start in the middle. Where Sweeney kills off the jeweller for his string of pearls. O.K.? I want Marshall up in the wings with Tom. The rest of you can sit and watch for the time being. Oh, and I want to get the shadow effect today. Stephen, can you come behind and work the lamp?'

'Your wish is my command.' Jumping on to the stage, Stephen disappeared behind the backcloth as Marshall and Tom started their scene. Marshall was at his most avaricious, leering at the pearls as Tom held them up and grinning at the audience with fearsome complicity.

'Just settle yourself in my chair, Mr Parrock,' he crooned, 'while I go and get – my razor.'

Grimacing horribly, he disappeared into the wings and, a second or two later, passed behind the backcloth. A square of gauze had been set into one of the painted windows and, as Marshall walked past it, it was lit up by a lamp from further back. He paused, his arm up-raised, holding something long and curved, his whole body set into an attitude of menace. The light cast his shadow sharply on to the gauze. Beside him, Jackus felt Ann shudder.

'He looks really dangerous,' she muttered.

Miss Lampeter was smiling appreciatively and nodding to herself, as if she had got the effect she wished to create.

The next moment, Marshall appeared on the stage, his hands raised and the long metal blade gleaming.

Miss Lampeter sat up straighter. 'Hang on a moment, Marshall,' she interrupted. 'What's that you've got there?'

Marshall smiled, a slow, satisfied smile and turned towards her, holding it out. 'It's a cut-throat razor. It was my grandfather's.'

Miss Lampeter came up to the front of the Hall and held out her hand for it. Turning it over and over, she ran her finger dubiously down the edge. 'I'm not sure it's quite the thing –'

'It's *exactly* the thing,' Marshall said firmly. 'Precisely the right kind of razor.'

'But is it safe?'

He gave her his most charming smile. 'Think I'll get carried away and cut a few throats with it?'

'No, of course not.' She laughed brightly.

'I'll take great care of it,' Marshall murmured. 'And it's really good. It puts me in the mood.'

He reached for it and, after a second or two, she nodded and dropped it into his hand. 'All right. But make sure it goes into the props. box after every rehearsal. It's not the sort of thing we want floating round the school. I don't know what Mr Garner would say.'

Marshall looked at her gravely. 'I'm sure he'll see it as part of the right historical effect. You've been trying really hard to get that, after all.'

She glanced quickly at him, as though she suspected he might be mocking her. But his face was innocent and earnest and, after a brief pause, she nodded. 'All right. Go on with the scene, then.'

Marshall made a flourish with the razor and moved round behind the chair. As he did so, Tom shrank away automatically and the audience laughed. But Jackus did not laugh. He was watching Marshall run his finger over and over the sharp blade, as if he were fascinated by it.

And the razor did seem to make a difference. After a scene or two, Jackus realized that everyone on the stage was treating Marshall with more caution than usual, adding an extra spice of tension. There was no temptation to laugh now. Sweeney Todd was armed and murderous.

There was another difference as well. Marshall himself seemed more urgent and reckless. Until that point, there had been something faintly amused about his acting, but now he was grim and threatening. When he seized Benny, to hale him off to the lunatic asylum, he gripped him not by the shoulder, as before, but by the ear, twisting it hard. Benny gave a loud cry of protest.

'Ouch! Marshall! What do you think you're doing? That *hurt*.'

'Don't make such a fuss,' Miss Lampeter said briskly. 'Marshall's right. It looks really good. We'll keep it like that.'

Rubbing his ear sulkily, Benny allowed himself to be haled off the stage. A moment later, he sat down beside Jackus, muttering under his breath.

'Don't know what's got into Marshall. I thought he was going to have my ear off.'

'I expect the Lamppost would like that even better.' Jackus grinned. 'Nothing like a bit of blood on the stage to set the atmosphere.'

'Huh!' Benny snorted. 'You volunteering to be the first real human pie, then?'

He turned his back and glared sullenly at the stage. He was not the only one. Everyone Marshall touched was wincing and frowning, but no one dared to protest again until the scene where Sweeney murdered Mrs Lovett. That had been rehearsed more than anything else. Ann was supposed to come in, sweating with fright, to say that she was not prepared to go on working with him. In the struggle that ensued, Marshall strangled her and then caught her by the shoulders and dragged her off the stage. At rehearsal after rehearsal, the audience had collapsed into laughter as Ann's large body thudded unconvincingly to the floor.

But, this time, no one even smiled. There was nothing funny about Ann's face as she confronted Marshall, her eyes glittering with fear. And when she slumped, Jackus thought, wildly, that she must actually be dead. Her whole body went slack and heavy, falling against Marshall, and he lowered her gently to the ground and stared down at her with gloating smugness.

'Oh *yes!*' Jackus heard Miss Lampeter whisper. She bent forwards, scribbling in her little notebook.

'You shculd not have quarrelled with me, Mrs Lovett, my dear,' crooned Marshall. 'Those who quarrel with Sweeney Todd always come to a bad end.'

Stooping, he clutched not at her shoulders but at her hair, grabbing a large handful of it and starting to tug her towards the wings. Ann screamed and sat up abruptly.

'Marshall! You're mad! You're pulling my hair out!'

Miss Lampeter sighed with impatience. 'Oh, Ann. Why did you have to stop? The scene's never gone so well.'

'But *he pulled my hair out*,' Ann said again, incredulously. 'Look.'

Sure enough, a strand of brown hair was dangling from Marshall's fingers. He dropped it on the ground and wiped them fastidiously. 'Shouldn't matter to you, should it? You're dead. You can't feel anything.'

'Why don't you strangle me properly and be done with it?' Ann rubbed her head, looking ready to cry, and Miss Lampeter gave an audible sigh.

'It won't do, I'm afraid, Marshall,' she said regretfully. 'It looks smashing, but I can't have you pulling Ann's hair out. Just think what her mother would say if she came home bald.'

Marshall shrugged. 'O.K. You're the producer. Shall I go on, then?'

Miss Lampeter looked at her watch. 'No, we probably ought to stop there. I hadn't realized it was so late. Pack up your things now. We'll meet again on Tuesday.'

She scooped up her own belongings and was already half-way down the Hall when Mandy gave a loud scream.

'Oh *no*! What am I going to do? My mother'll *kill* me!'

Miss Lampeter paused and looked back. 'What's the matter *now*, Mandy?' Her voice sounded weary.

'It's my gold cross and chain,' Mandy wailed. 'I took it off, because the catch was loose, and I put it in my bag. Now it's gone.'

'Oh, you silly girl!' Her tiredness making her sharp, Miss Lampeter came striding back towards them. 'You shouldn't bring things like that to school. It always causes trouble.'

'But I had it for my Confirmation. I was supposed to wear it *always*. I don't know what Mummy will do if she knows it's lost.' Mandy began to cry, her pretty face blotching into a mess of red and white.

'All right, all right. Don't get so upset.' Rather awkwardly, Miss Lampeter put an arm round her. 'Come on, everyone, have a good look. It must be somewhere.'

Grumbling, people started to rummage through their bags and pockets, to crawl around the floor peering under the chairs. Jackus joined in, but with an odd feeling of unreality. Was it possible that no one except him realized that this had happened before? He kept expecting a gold chain to come whizzing over his shoulder from nowhere.

But it did not happen. And no one yelled, 'Here it is!' The only voice to be heard was Mandy's, choking with sobs.

'I suppose I'd better go through your things myself,' Miss Lampeter said, a trifle tartly. 'I expect it's really in your bag after all.'

She seized it and sorted through for a moment or two. Then she paused and looked up at them all, her lips tight. 'I wouldn't like to think,' she said slowly, 'that anyone had taken it on purpose. It's probably fairly valuable, isn't it, Mandy?'

'It's s – solid gold,' Mandy stuttered. 'My godfather bought it for me and – oh!'

She wailed again and Miss Lampeter glanced round helplessly.

'Shall I take her down to the cloakroom?' said Ann, stepping into the breach. 'To have a drink and wash her face?'

Miss Lampeter nodded and, when the two of them had gone, turned to the rest of the cast. Her face was unusually grave and severe.

'Look, I don't know what's happened to it, but you must all realize that if it doesn't turn up I shall have to report the whole matter to Mr Garner. We can't simply ignore something like this. He may decide to call in the police.'

Jackus saw Marshall's eyes flick quickly towards him and then flick away again. Irritated, he pulled a face as Miss Lampeter went on speaking.

'It's not very pleasant to think we may have a thief in the cast. But that's what it's beginning to look like. Is there anything anyone would like to say?'

She stopped, looking round, but no one made a sound and at last she shrugged. 'Very well. But I suggest that, if anyone knows anything about it he or she takes steps to see that Mandy gets her property back. Now you'd all better go home.'

She started off down the Hall again and, watching her, Jackus was aware that it had been the last straw. Her shoulders drooped, and she did not walk with any of her customary bounce. He had not seen it before, but now he understood how much the play was taking out of her.

He was so busy staring at her that he did not hear Marshall come up behind him. Suddenly, in his ear, a humorous voice spoke.

'I'm surprised at you, Jacko.'

Turning, Jackus gazed into a pair of curious eyes.

'What are you on about, Marshall?'

Marshall glanced round quickly, to make sure that he was not overheard. Then he muttered coldly, 'I thought you were a reformed character.'

'What –?' began Jackus. But Marshall was already on his way down the Hall, without so much as a glance over his shoulder. Jackus shivered suddenly. Then he made himself smile. It must have been a joke. One of Marshall's sharp, deadpan jokes.

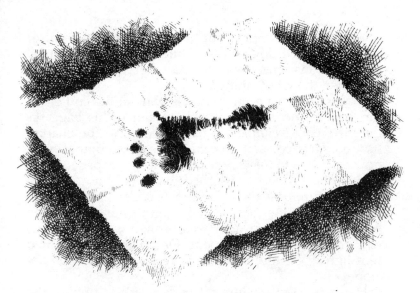

Chapter 8

It was safe here, Jackus decided. The rest of the school was full of people milling about in a bored way, watching the water stream down the windows. But here, in the little cubby hole behind the Art room, things were quiet. Hardly anyone came up here in the dinner hour. It would be the ideal place to think.

And he needed time to think. Settling himself on a stool among the dirty paint pots and stacks of paper, he rested his chin in his hands and stared out at the grey sky and the driving rain. What he wanted was a link, something that would make sense of all the peculiar things that had been happening. Because he had no doubt that they were peculiar. Chewed sandwiches, books flying through the air, things disappearing – there seemed to have been a long succession of incidents, each one odder than the last, but he could not fit them together to make a pattern. Thoughtfully, he reached into his pocket and drew out a piece of paper,

73

carefully folded. Spreading it on the table in front of him, he looked down at the blurred prints of the small feet. His evidence that he had not gone mad and imagined the whole thing.

He brooded for a moment, chewing at the end of his forefinger, and then took a book out of his bag. *The Complete Ghost-Hunter's Guide.* Turning to the chapter headed *Poltergeists*, he began to read, stopping every now and then to make notes on a sheet of paper.

He was so deeply absorbed that he did not notice a plump figure pass the door. It paused, stepping back to peer through the glass panel. He did not notice even when the half open door was pushed wider and Ann walked quietly into the room. Only when she touched him lightly on the arm did he realize that he was not alone.

He came back to himself with a jerk, dropping the book loudly on the floor.

'Sorry,' Ann said. 'I didn't mean to startle you. I just –' Bending quickly, she picked up the book. But instead of giving it back to him, she turned it over in her hands, reading the title. 'Yes,' she said softly. 'I thought that was what I saw.'

'Give that to me!' Jackus said roughly. He snatched the book out of her hands, jamming it into his bag.

But Ann was glancing round the room. When she saw the piece of paper spread on the table, she drew in her breath sharply and looked back at him.

'Why have you got that? And why are you reading that book?'

'None of your business. What are you doing snooping round, anyway? I came here for a bit of peace and quiet.'

'I wasn't snooping.' She flushed. 'I just wondered what you were doing in here. And when I saw what you were reading –' She sat down on a stool beside him and gazed at him earnestly. 'Jackus – have you noticed anything odd going on? At rehearsals, I mean?'

'Don't know what you're talking about.' But his eyes were on her face, trying to work out whether she would make fun of him.

She did not look as though she were ready to laugh. Her round face was distressed and she rubbed miserably at her forehead, searching for words to explain what she meant.

'Well, there was my sandwiches, to start with. I was sure it was you, but I've thought and thought about it since, and I don't see how it can have been.'

'Brilliant!' Jackus grunted. 'I told you it wasn't me.'

'But who else could it have been?' Without waiting for his answer, she plunged on. 'Then there was that pencil that someone threw at Miss Lampeter. That wasn't you either. But there wasn't anyone else there. I *saw*.'

'Good at seeing things, aren't you?' Jackus jeered, a little uncertainly. 'Perhaps you need your head examined.'

'Don't!' She shuddered. 'That's not funny. I've thought of that, too. And I've been wondering and wondering if I'm going mad. When I saw that footprint on the Art room floor, and you said it was nothing, I felt –' Her lip trembled briefly, and then she snatched at the paper on the table. 'But I *didn't* imagine that, did I? You saw it, too. And you did think it was important, or why have you got this with the same footprint on it? The one with the toe missing?' Triumphantly, she faced him, her cheeks pink. 'And why are you reading that book?'

He shrugged.

'Oh, Jackus, you've got to tell me. Don't be so mean. Can't you see how worried I am?'

He considered. 'What happens if I do talk to you? More silly ghost jokes? More people putting me off while I'm acting?'

She went even pinker. 'I'm sorry about that. But it wasn't my idea. It was – well, it doesn't matter. And I won't do it again. I need to know what you think.'

Jackus looked at her serious face, at her hands, trembling on the table. Suddenly, he made up his mind. 'It's odder than you think,' he said slowly. 'There's other things that have happened. Things you haven't mentioned.'

Not really expecting her to believe him, he began to explain. When he told her about the bare arm that had pushed him, struggling to get at the bread, she shivered and rubbed her own arm absent-mindedly. But when he described how the books in the library had flown from their shelves, her eyes grew wide.

'Weren't you terrified?'

'Dunno. Didn't have time to think about it, really. Not in the beginning. And then, after I was stopped from going up to the end of the room, I knew, somehow, that whatever it was wasn't trying to hurt *me*.'

'Whatever it was,' Ann repeated softly. 'So you do think it's something?'

'Can't make my mind up. All the things that have happened, they're all strange, but they don't seem to hang together. Why should they suddenly start happening now?'

'Oh, but that's *obvious*.' Ann was so eager to explain that she lost her frightened look and leaned forwards, excited. 'It's the play. Everything that's happened has had something to do with the play. Even in the library, you were supposed to be learning your lines.'

'I wasn't even doing that,' Jackus said, understanding all at once. 'I was saying some of Marshall's speeches out loud. Some of the really cruel, horrible ones.'

'There you are, then!' She sat back and banged her fist on the table. 'That makes sense, doesn't it? It's such a violent, beastly play. It's making violent, beastly things happen.'

Jackus looked unimpressed. 'You could be right. But it doesn't get us any further, does it? I mean, we don't understand things any better.'

'There must be a way we can find out more,' Ann said

impatiently. 'I don't think I would be so scared if I knew what was going on.'

Jackus tapped a finger lightly on the cover of the book that lay in his lap. 'There is a way,' he said at last. 'That's what I was reading about when you came in. How to investigate ghosts. But I don't know if we dare.'

'Go on. What do we have to do?'

'We have to raise the ghost. Do something on purpose to make it come.'

Gulping, Ann looked down at her fingers, winding them together. 'You really think – that it is a ghost?'

'It behaves like a ghost. Or a poltergeist, anyway. Flinging things about the place.'

'And d'you think it would be safe? To do something like that?'

That was precisely what Jackus had been wondering when she interrupted him. But now, looking at her scared face, he felt braver. At least his teeth were not chattering. 'How should I know if it's safe? But it's the only way we're going to find anything out. Or are you too frightened to try?'

'I'm not a coward!' Her face was pale now, but she said the words stubbornly and, for the first time, Jackus felt a grudging twinge of liking for her. She might not be the person you'd choose to have in a tight corner, but she was certainly game. He had to admit that.

'I've got a sort of idea of what we could do,' he muttered. 'To try and discover if it's connected with the school, or if it's just the play that's causing everything. Are you on? Will you help?'

'Yes.' She said it rather too loudly, as if to assure herself of her own determination. And before she could even waver, the bell sounded for the beginning of afternoon school. Jackus got to his feet.

'Right then. You'd better come to my house. Saturday. About eleven in the morning.'

She nodded and followed him towards the door, not

77

speaking. As they stepped out into the corridor, a sarcastic voice came floating down from the far end.

'My, my, what have you two been up to, then? Lurking around like that?'

They whirled round and met Marshall's grin. He came down the passage towards them. 'Thought you had better taste, Little Orphant Annie,' he said lightly. 'Or can't you get anyone except Jackus to spend the dinner hour with you? Don't you think you should pick on someone more your own size?'

Ann's face was flaming, and Jackus felt a sudden spasm of annoyance. 'What's got into you, Marshall? You've been getting more and more rotten, these last few weeks.'

But Marshall did not pay him any attention. He was gazing fixedly at Ann. 'Go on,' he said, almost gently. 'Tell. What have you been up to?'

Abruptly, Jackus realized what a fool she could make him look. She only had to turn the whole thing into a joke and pretend she had been teasing him and she would be on Marshall's side. And that was what she was sure to want. Everyone knew she was soft on Marshall. He glanced angrily at her, waiting for the laughter to start.

And she did laugh. But it was a short, scornful laugh. 'You are an idiot, Marshall. What do you *think* we've been doing?'

'I can't imagine.'

Jackus held his breath.

'We've been rehearsing, of course,' Ann said firmly. 'Jackus is so lousy that someone's got to give him a bit of practice or he'll get chucked out of the play. And then we'll all be in trouble, because there isn't anyone else.'

'Such a public-spirited girl,' Marshall murmured. 'You put us all to shame.'

'That wouldn't do you any harm,' retorted Ann. She glared at him and, after a second, he shrugged and walked off down the corridor.

'Thanks,' Jackus said quietly. 'I thought you might – well, thanks, anyway.'

'I wasn't going to tell *him*, was I? I thought I might, once, but after the last rehearsal I changed my mind. He was really peculiar. Pulling my hair like that!'

'I think he just got carried away by the acting,' Jackus said. 'He likes to do things properly, you know.'

Ann raised her eyebrows. 'Why are you making excuses for him? Anyone would think he was a friend of yours.'

'But he is,' Jackus said, without thinking. He was surprised she did not know it. 'We've always been friends.'

'You? And Marshall?' She looked completely baffled. 'What do you mean? You never go round together. You never do things together.'

'I suppose we don't,' Jackus said slowly. He had not thought of it like that. 'But Marshall's always been busy at school. Joining in things. I'm not a joiner.'

'So how can you be friends?' Ann persisted. There seemed to be something about the idea that intrigued her.

'Well, it started with his mum and my mum. Being best friends. I always played with him when I was a little kid and, outside school, we do things together.' For an instant he was briefly amused, wondering what she would say if she knew some of the things he and Marshall had done.

But she was not thinking about that. She was staring at him in a puzzled way. 'And you're *really* friends? I mean – you like each other?'

That question again. That silly question. Jackus was growing annoyed at her prodding. Lightly, he said, '"Course I like him. He's bigger than me and cleverer than me and everyone thinks he's fantastic and I'm stupid. Gives me someone to look up to, doesn't it?'

To his surprise, Ann suddenly looked embarrassed. She put a hand on his arm. 'I don't think he's fantastic and you're stupid,' she said awkwardly. 'I think you're much nicer than he is.'

'Coo, thanks,' Jackus said, even more embarrassed than she was. 'That's really made my day, hasn't it?'

'Well, there's no need to be horrid about it.' She tossed her head. 'I was just trying to be friendly.'

She started to walk away from him, in a huff, and Jackus felt mean. But not quite mean enough to apologize. Instead, he called after her, 'Don't forget. Saturday.'

She nodded, and then turned and gave him a friendly grin before she ran off. As Jackus followed her, it suddenly came to him that he had an ally in the school. Someone on his side. He could not remember when things had been like that before and he found it a strange feeling. But rather a comfortable one.

Wednesday, 2nd December

'. . . so I'm not mad after all! Whatever it is, it's not me. I'M NOT MAD!!!

When Jackus told me that he had noticed all the queer things too, I was so relieved that I could have hugged him. Imagine how surprised he would have been! He'd probably have had a fit. Somehow, I don't think he expects people to be nice to him and like him. That's why he skulks round the school in that horrid way, sneering at everything. He's getting at people before they have a chance to get at him. And it's certainly always made me think how beastly he was. But now I've talked to him, I think he might really be O.K.

I *hope* he's O.K., anyway. Because I've agreed to do something with him. Something that might turn out to be rather dangerous. When I think about it, it makes me scared, but if we don't do it, we'll just have to sit back and let peculiar things go on happening. And that would be MUCH WORSE.

At least I'm fairly sure he won't turn round and start to make fun of me, and pretend it was all a joke. He was quite nice and serious when we were talking about it, and afterwards, when we bumped into Marshall, he didn't rat, although I thought he might for one horrible moment, when Marshall was being rotten.

But he did say something odd when Marshall had gone. He said they were friends. That they went round together out of school. And I think he was telling the truth. I wanted to ask him, if that was true, why Marshall dreamed up that rotten plan about the ghost jokes to get him into trouble. Because it *was* rotten. I can see that now. I felt really squirmy to think I'd had anything to do with it. But I didn't tell after all. It would only have made trouble, and I couldn't bear any more trouble just now, what with the play being so queer and Marshall being so foul.

Because he is absolutely one hundred per cent certified FOUL. I've never liked him really. He thinks he's the most wonderful person in the school and everyone else is rubbish.

If I had a wax model of him, I'd stick a great big hat-pin right through its heart.

And I wouldn't take it out again.'

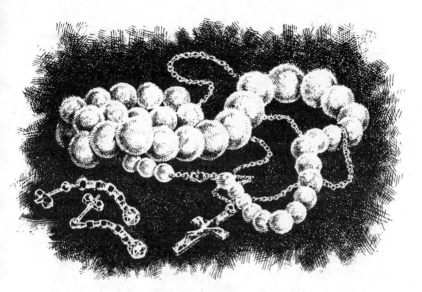

Chapter 9

'But do you not remember him at all, Mr Todd?' Mandy said pathetically. 'A sailor that came in a week ago for a shave? We were to have been married tomorrow, but he has not been seen since, and I am in despair!' She gazed across at Marshall, her big eyes pleading.

'Bless you, miss.' Marshall gave a faint snigger. 'Do you imagine I can remember every face I shave? They're all chins to me, my dear.'

'Then I must pine alone and never marry, since he is lost.' Choking back a sob, Mandy walked off at the side of the stage, while Miss Lampeter nodded her approval, and Marshall slowly opened a small box that stood on a table beside him.

'A sailor that came in a week ago,' he murmured, pulling something out of the box. 'Oh no, I have not forgotten him.' He chuckled. 'These were fine diamond ear-rings he brought from the Indies to decorate your pretty ears, miss. Too fine for old Sweeney to let them go.' Marshall dangled them in his hand, letting them swing so that the light glinted on the faceted glass. 'And when I see something that I want – I take it!'

With a sinister leer at the audience, he dropped the ear-rings back into the box again and stropped the razor against his left hand.

'That's coming along nicely.' Miss Lampeter stood up. 'If anything, Marshall, you need to be a bit more obsequious when you're talking to her. To make the contrast stronger when she's gone. But I think that will do for today.' She shuffled her papers together and looked at them gravely. 'Now, there's something else, I'm afraid. This business of the cross and chain. I don't suppose you've found it, have you, Mandy?'

Mandy shook her head. 'My Mum says you ought to do something about it. She was ever so annoyed.'

'I'm sure she was,' Miss Lampeter said with a sigh. 'It's most unpleasant. Well, I have done something. I've told Mr Garner about it, and he wants to see you all in his room now.'

There were suppressed groans from some of the cast, but Stephen said, 'Good. I'll tell him about my money, then.'

'What?' Miss Lampeter looked startled. 'You've lost something as well?'

'Pound note,' said Stephen, blinking slightly as they all turned to look at him. 'I didn't realize until I got home after the last rehearsal, but I'm sure it was in my pocket.'

'Oh, this is too bad!' Miss Lampeter looked fretful. 'Is there anyone else? We might as well get it all out in the open before we go and see Mr Garner.'

'There's my calculator,' Benny said slowly. 'I thought I might have left it in the Maths room, but I've looked everywhere and I can't find it.'

As if he had pulled out some kind of stopper, people began to add things. There was nothing very valuable – a few coins, a bag of sweets, a fancy handkerchief – but the list grew longer and longer. After a moment or two, Miss Lampeter took out her notebook and wrote them down and when, at last, people stopped talking she rubbed her face and said pettishly, 'Oh, it's too bad! As if doing this play wasn't difficult enough, without something like this happening! Well, we'll all go and tell Mr Garner about it. Perhaps he'll be able to sort it out.' In a thoroughly bad temper, she stamped away down the Hall, leaving them to straggle after her. They followed quietly, in twos and threes, glancing suspiciously at each other. It seemed to Jackus that most of the eyes fixed on him, but he told himself that it was just his imagination. After all, they did not have any proper reason for thinking it was him. It was not as if they *knew*.

Ann sidled up to his elbow and prodded him. 'I don't like it,' she whispered.

'Don't suppose anyone else is over the moon,' Jackus grunted.

'No, I mean, do you think it's anything to do with –' She prodded him again. 'You know.'

He shrugged. 'How can I tell? Perhaps we'll find out on Saturday.' He suddenly realized that Marshall had turned and was looking back at them. 'Shut up now,' he added in a low voice. 'I don't want to talk about it here.'

By the time they reached Old Garner's room, Miss Lampeter was already in the middle of her explanation, wagging the list of stolen items under the headmaster's nose.

'It's most unpleasant!' she was saying hotly. 'There must be a petty pilferer in the cast, and I don't see how we can go on with everyone wondering if someone else is getting ready to steal. You'll have to do something about it!' Her voice had risen, almost to a shout, and as she stopped she actually stamped her foot on the floor.

'I'm sure it's most upsetting for you,' Mr Garner said calmly. 'Why don't you sit down and let me talk to them?'

He waited while she bumped down into a chair and then looked round at the members of the cast ranged in front of him.

'I'm sure you're all aware of what Miss Lampeter has just been telling me,' he began. 'And I must tell you that I find it most disturbing.'

'The disturbingfulness is terrific,' muttered Stephen. He subsided into silence as Old Garner glared at him.

'May I ask, first of all,' the headmaster went on, 'who has *not* had anything stolen? Perhaps you would raise your hands.'

Slowly, Jackus, put his hand up, aware that Miss Lampeter glanced sharply at him as he did so. He put it up higher, defiantly. At least he was not the only one. Ann and Marshall had their hands up too.

'I see.' Mr Garner looked down at his fingers. 'You three must realize that you are the prime suspects.'

'I don't see why,' Marshall drawled, not sounding in the least bothered. 'If I had stolen all those things, I'd make sure I'd lost something myself as well. To avoid suspicion.'

The headmaster smiled wryly. 'I'm sure you would, Marshall. Nevertheless, I take your point. Very well, then I shall make my appeal to all of you.' His eyes flicked round the room, resting on one face after another. Some people shuffled awkwardly, and Benny Harris went pink. But Jackus knew that did not mean anything. Benny was always going pink.

'A large number of things has been taken,' Mr Garner said, putting his fingertips together. 'Most of them are not very valuable in themselves, but that is not what I am concerned with. What distresses me is that one of you has clearly been stealing from the rest of the cast in a systematic way. And I cannot let that pass. What I should like, is for the thief to confess now, and restore the things that have been taken. If he – or she – does that, I think we might forget the whole matter.'

Fat chance, thought Jackus scornfully. Even Old

Garner could not believe that would happen. But he stared at them all for a full minute, giving the opportunity for the guilty person to speak. Most of the cast kept their eyes firmly fixed on the ground, but Jackus suddenly became aware that Miss Lampeter was gazing at him. He glared back, until she shifted her eyes.

'Very well,' Mr Garner said at last. 'If the thief is not prepared to confess, I have no choice but to make a threat. Miss Lampeter has said that the play cannot go on in this atmosphere, and I agree with her. Therefore, I shall demand that the stolen property be restored by the dress rehearsal next Wednesday. If not –' he took a deep breath '– I shall cancel all three performances of the play.'

Miss Lampeter gasped, and looked round at him. 'But you can't –'

He lifted a hand, signalling to her to be silent. 'I know this is a rather drastic solution, but it seems to me to be the only one. All of you have, I know, put a great deal of effort and enthusiasm into this production. I hope that you care enough about it to want to save it. If every one of you feels like that, we shall have no problem, shall we?'

'Oh, please,' Miss Lampeter broke in, 'whoever it is, *please* give the things back somehow. It's all so stupid.'

She looked straight at Jackus as she said it, and he found himself growing angry. That was all it meant, was it? All her understanding smiles and her little pats on the arm did not stop her picking on him when something like this came up. He stood stubbornly still when Mr Garner dismissed them, waiting until the room was clear except for the two teachers. Then he stepped up to the desk.

'Got something to say, sir.' The sound of Miss Lampeter's quick, relieved breath did not improve his temper.

'Go on, Jackus.' Old Garner's face was blank. No chance of guessing what he thought.

'I just wanted to say – it wasn't me, sir. I didn't take those things.'

Miss Lampeter made an impatient movement, but Mr Garner smiled slowly, amusement on his sharp face. 'It may surprise you, Jackus, but I never supposed that you did. I don't think that sneaking things out of people's bags is quite your style.'

For a moment, Jackus was nonplussed. In his confusion, he said, more rudely than he had intended, 'But *she* does, sir. I saw her looking at me when you were talking. She's sure it's me.'

'Oh, Colin, I didn't mean –' Miss Lampeter blushed and fell silent. Old Garner's smile disappeared.

'I dislike hearing you refer to Miss Lampeter in that impertinent way,' he said sternly. 'You will apologize.'

'She's the one who ought to apologize,' Jackus said sulkily. 'Staring at me like that. I bet everyone could tell what she was thinking.'

'That's enough.' Abruptly, Mr Garner stood up. 'I have been very considerate to you, Jackus. I have not told anyone except Miss Lampeter about your earlier – transgression – and I do not intend to do so. But you must realize that one of the penalties of theft is that you will be regarded with a certain doubt. If Miss Lampeter has wondered about you, she is quite justified in doing so. And you will apologize for being rude to her.'

'Sorry, Miss Lampeter,' Jackus said, grudgingly. 'But it wasn't me.'

'That's all right, Colin.' She gave him a wide, forced smile. 'If Mr Garner's happy to take your word, then I am too. I was just rather upset about the play, that's all.'

'I'm sure everyone understood that.' Old Garner inclined his head. 'Now, if that is all, Jackus, I suggest you go home.'

'Yes sir.'

'Unless you have any idea who is behind these thefts.'

It was said almost casually, catching him on the hop as he turned to go. For a fatal second, Jackus hesitated, and then he realized that he had waited too long to give a simple denial.

'I don't know who it is, no,' he said unconvincingly.

'And do you know anything about it?' The head-master's response was very quick. 'Because, if you do, you would be wise to tell me about it.'

'No, sir. Nothing definite.' He made himself turn and look back at them as innocently as he could. 'But if I do find out anything – anything definite – I'll tell you.'

'Very good.' Mr Garner gave a nod of dismissal and, as he closed the door behind him, Jackus heard Miss Lampeter launch into a flood of excited speech. He grinned to himself. Trying to save her precious play, was she? She ought to know better than that. Nothing would shift Old Garner once he had made up his mind.

As he went into the cloakroom, Jackus expected to find it empty. But it was not. Marshall was sitting underneath the pegs, with his long legs stretched out on the bench.

'Hallo, Little Colin,' he said annoyingly. 'Been saving your soul, have you?'

'I don't know what you mean.' Jackus unhooked his coat from the peg.

'Oh no?' Marshall's eyebrows went up in disbelief. 'I assumed that was why you stayed behind. To get your confession off your chest.'

'Don't be thick.' Amiably, Jackus aimed a punch at him. 'You know it wasn't me.'

'Do I?'

It was said very lightly, without any kind of emphasis, but the meaning was clear. Jackus looked at him incredulously.

'You're not serious? You don't really think it was me?'

'Why not? You're the obvious person. You're the only one of us who is in the play against his will, and you have been known to put your fingers on things that don't belong to you. Haven't you?'

'But that was quite different. You know it was different.'

'Do I?' Marshall said again.

'Yes you do.' Feeling hurt, Jackus started to shout. 'That other thing was just for a giggle. You know all about it. I don't know how you, of all people, could suggest –'

'*This* is what I'm suggesting.' Marshall uncoiled himself from the seat and stood over Jackus, tall and threatening. 'I'm suggesting that things will get very unpleasant for you if you allow Old Garner to stop the play. Because *I* want the play to go on. And what I want, I get, Little Colin. Just you remember that. So you'd better go home and think over your sins and decide what you're going to do.'

'You're crazy!' Jackus yelled at him. 'Even Old Garner knows I didn't take those things. That's what I stayed behind to tell him. And *he* believed me. So why don't you? You're supposed to be my friend, after all.'

'Of course. How silly of me. I forgot.' With an ironic grin, Marshall turned and walked out of the cloakroom. At the door, he paused. 'Just don't forget what *I've* said. I was quite serious. If the play's stopped, you'll be sorry.'

He swaggered off, and Jackus stood for a minute, his hands trembling. Marshall believed he had done it. He really believed it. That was amazing. But even more amazing was how upset he felt at Marshall's accusation.

Chapter 10

'We're really going to do it, then?' Ann said in a small voice. She looked round the small cluttered garage. 'Here?'

'It's the best place,' Jackus said, busy carrying things through the door. 'Dad won't be back until the evening, and if Mum comes in from shopping, she'll just think I've gone out. So no one will disturb us.'

'No one *real*.' Ann shuddered. Jackus looked at her curiously. For the past few days, he had been planning what they would do, but his thoughts had all been practical. He had had no time to worry about anything else. Now he saw that Ann was afraid.

'Want to back out?'

'Of course not.' She shook her head defiantly. 'I'm not scared. It's just – Well, it would be nicer if I knew what to expect.'

'If we knew what to expect, we wouldn't need to do it

at all,' Jackus said reasonably. 'Here, sit down up this end while I get ready.'

'But it's filthy.' She wrinkled her nose as she looked down at the dusty concrete. 'I can't sit on that.'

'I never pretended it was the Ritz,' Jackus said, a little crossly. He spread out a piece of newspaper. 'There you are, madam. Your throne.'

She sat down, giggling nervously, and watched while he made his preparations. First, he laid another sheet of newspaper on the ground in the middle of the garage. On it, he placed the things he had brought. The back end of a loaf of bread. A neatly folded handkerchief. A few small coins. And half a lamb chop that his mother had not felt like finishing. Ann pulled a face as he set it down.

'That's enough to put anyone off. D'you think they – it – will pay any attention to that lot of rubbish?'

'Could do.' Jackus was preoccupied. He had arranged the things neatly and was busy drawing round them, on to the newspaper, to mark their positions exactly.

'You need something more valuable than that,' Ann persisted. 'A watch or something.'

'Haven't got one.'

'Oh, have mine.' She took it off and held it out to him. Putting it down on the newspaper, Jackus looked at it.

'Do you really want to risk it?'

'It'll be all right.' She giggled again. 'I don't actually think anything's going to happen.'

I bet, Jackus thought, looking sideways at her tense face. But he did not say anything. Instead, he took the packet of flour out of his plastic carrier bag. When Ann saw it, she snorted.

'What's that for? Think ghosts makes cakes, do you?'

'Shut up. You'll see.' Starting at the far end of the garage, he began to sprinkle the flour evenly over the floor and the things on the newspaper, walking backwards to avoid leaving marks on it. When he had finished, there was an even, white coating all over the dirty concrete.

92

'Oh, *very* clever.' Ann settled herself more comfortably. 'What do we do now?'

'Now,' Jackus said, 'I put out the light.' As he reached up for the switch, he saw her move, as if to protest, and he grinned. 'It's O.K. I've got a candle. I won't leave you in the dark.' He flicked the switch and fumbled in his pocket for the matches. Momentarily, he could see nothing. The garage had no window, and the thin streak of light which came in under the door reached only a few inches across the floor. There was a scrape as he struck the match against the side of the box and held it to the candle wick. The tip of the wick grew bright, wavered once or twice and then started to give a steady flame. Bending forwards carefully, so that he should not mark the flour, he set the candle down a little way in front of them and then sat on the newspaper beside Ann.

'Aren't *you* scared?' she murmured. 'Just a little bit?'

'Not really.' He considered. 'I think that if we're right, if anything – or anyone – comes to pinch those things, *it'll* be more scared than we are. That's the thing about stealing –'

Beside him, he felt her stir with interest, but he went on for a moment, pursuing his own thought. 'Imagine it. There's your heart thudding, so loud you think everyone for miles around can hear. And it's impossible to breathe. Like suffocating. And all the time he – if there's anyone with you – he's glaring at you, as if he'd throttle you if you made a sound.'

'How did you –?' Ann stopped, as if she had decided not to say it. Jackus chuckled sourly.

'No, it wasn't me. I've just got a good imagination.'

She shivered, her teeth chattering audibly. 'I wish I hadn't. That's what makes it all so terrifying.'

'Let's start. Then you won't feel so bad.' No sense in letting her work herself up. 'I think we need a bit of the play. I'm sure you were right about that. Let's do our bit together.'

She gulped, but when she spoke, her voice was steady, the familiar deep growl of Mrs Lovett.

'Be off with you! I can't be bothered with beggars.'

'I'm not a beggar, mum.' Purposely, Jackus made his words as cheerful as possible, to try and reassure her. 'I'm just a poor lad as is looking for an honest job of work.'

All the time his ears were alert, waiting for some sound, some movement from the middle of the garage. But there was nothing. Except that the whole place seemed to have got much colder. He rubbed his hands together to warm them.

'Have you got anyone to give you a good character?' Ann said sharply. As she spoke, the candle flame dipped and guttered, and he heard her catch her breath. Putting out his hand, he touched her on the shoulder to stop her saying anything, and went on with his lines.

'Bless you, mum, I don't need a character. Just look in my eyes. What do you see? Honesty! Pure honesty! I –'

The candle flame sputtered again and then, quite suddenly, went out. Ann's shoulder grew tense under his fingers and, from the far end of the garage, came a sound. A very small sound. Like indistinguishable whispering. He felt Ann move and her hand gripped at him.

'Ssh!' he said under his breath. 'Don't spoil it. Go on with the scene.' His own throat was so dry that he could hardly talk. Ann stirred and began to speak again, but this time it was in her own voice, high and strained.

'Please,' she said softly. 'It's all right.'

'What?' He prodded at her, but she did not take any notice of him. After a second or two, she spoke again.

'It's all right. You can have the things. We won't hurt you.'

'Ann!' He fumbled in his pockets, trying to find the matches so that he could light the candle again and see her face. But he must have dropped them. Beside him, he could hear her breathe harder in the dark. Suddenly, she dropped forwards on to her hands and knees, her voice

94

pleading. 'It's all right. It's not a trap. You don't have to be afraid of us. We –'

Then, in a single movement, she jerked back, knocking Jackus's arm sideways. Her voice came out in a single yell.

'Oh!'

And she began to sob, taking great shuddering breaths, her whole body rocking backwards and forwards.

'Ann! Don't!' Jackus tried to shake her, but the force of her sobbing threw his hands aside. Thoroughly frightened now, he jumped up and lunged at the light switch. It clicked down, filling the whole garage with bright, prosaic light.

The floor was in a chaotic mess. The flour had been trampled everywhere by innumerable small footprints and everything that had lain on the newspaper in the middle was tossed wildly about. He could see the lamb chop, horribly gnawed, lying near the candlestick. The loaf of bread, crumbled into fragments, was scattered in different corners and the handkerchief and the coins were in a higgledy-piggledy heap up by one wall. Of Ann's watch, there was no sign at all.

Ann herself was hunched forward, her arms clutched round her head, rocking backwards and forwards. Awkwardly, Jackus knelt down and put an arm round her.

'It's all right. They've gone. You're safe now.'

'Oh no! You don't understand.' She raised a blotchy face, messy and smeared where she had rubbed it with her dirty hands. 'Didn't you feel it?'

'Feel what?' He patted her back. 'Honestly, I didn't feel a thing. Just heard a few noises. And you talking. What were you on about?'

'I – oh –' She looked as though she might start to wail again, so he fished out his handkerchief and held it under her nose.

'Here. Clean your face up a bit. You look awful.'

'Thanks.' She blew her nose hard and then licked a corner of the handkerchief and began to scrub away at her face with it. As she did, she gradually calmed, and Jackus took his arm away and looked at her.

'O.K. now? Can you explain?'

'You *really* didn't feel it?' Her eyes were wide with disbelief. 'It was thick in the air. Like – like a fog.'

'It's no good saying that. I don't know what you mean.'

'Well.' She sniffed, blinked and sat up straighter. 'First of all I felt cold. Didn't you?' He nodded. 'And after that, I began to feel hungry. Desperately hungry, as though I would die if I didn't get something to eat soon. And that was queer, because I've only just had my breakfast. And then I realized – it wasn't *me* that was hungry.' She frowned, trying to find a way to explain. 'You remember that day we were in the Geography room? When I said I could hear Mrs Lovett's voice in my head, and it was me and not me? Well, it was like that. As though the air was full of hunger, and I was feeling it from outside. And then – and then the fear started.'

'You were afraid of ghosts?'

'No, *no*,' she said impatiently. 'It was from outside. Like the hunger. A horrible, creepy sort of terror. As if nowhere was quite safe enough.'

'Oh, I *see*.' At last something made some sense. 'That's why you called out to tell them not to be afraid. It was them.'

She nodded. 'I thought they were afraid of *us*.' She laughed abruptly, without any humour. 'I hadn't got the idea at all, you see. And then it came suddenly. Like a picture in my head. Like when I knew what Mrs Lovett looked like.'

'What picture?' He could happily have shaken her. 'Go on. Tell it properly.'

She narrowed her eyes, as though she could still see it. 'It was a horrible old man. Creeping, creeping. Always behind them. Always watching and taking and threatening. *Evil.*' She dropped her head into her hands and muttered, muffled through her fingers, 'I'm going mad, aren't I? You didn't feel a thing. I'm just going off my head and imagining it all.'

'Don't be soft,' he said gruffly. 'You were sitting beside me all the time. I had my hand on your shoulder. So how could you have done *that*?' He gestured widely, indicating the mess in the garage.

Ann looked up and, as he saw her face, he realized that she had not even noticed the way things were tumbled about. What she had felt must have been so real to her that she had forgotten about everything else. As she stared, her expression lightened, changing to one of immense relief.

'So it wasn't just me.'

'That's what I told you.'

Slowly she got up and began to walk round the garage, looking at the things on the floor.

'Your watch seems to have gone, I'm afraid,' Jackus said apologetically. 'Can't see it anywhere. Will your Mum be furious?'

Ann made an impatient gesture, as if he were saying something of no importance. Stooping down, she picked up the chewed lamb chop and turned it over in her hands, biting her bottom lip. 'I can't imagine,' she said, as if she were strangling, 'I can't imagine being as hungry as that.'

'Hungry ghosts?' Jackus grinned. 'Sounds mad, doesn't it?'

But she did not smile. 'Suppose,' she murmured, 'that what causes ghosts is strong feelings. If you concentrate all your energy into one kind of feeling, it might make a kind of print, that goes on after you're dead. Perhaps it stops you knowing you're dead even. The strength keeps on and on, even without a body. Then wouldn't it be terrible –'

She paused for a moment, as if what she was thinking was too horrible to say all at once.

'Yes?' Jackus prompted her.

'Well – don't you see what that would mean? It would mean that all those ghosts – and they were children, they must have been, because their feet are so small – all of them concentrated their energies into hunger. Hunger and fear. That was their *life*. Oh, imagine. How awful!'

But he could not imagine. It was a bit like being blind. She was explaining something that meant nothing to him. Half joking, he said, 'It was a long time ago, anyway. They're dead, aren't they?'

She whirled round, furiously. 'You wouldn't say that if you'd felt how afraid they are. Not like dead people at all. They're hungry and afraid and trapped and – oh, there must be some way we can save them from *him*. Somehow we must stop him and get rid of him and –'

He tried to calm her down by being rational. 'Don't be daft. He's nothing. Nowhere. How can we do anything to him?'

'Of course we can. Don't you understand *anything*? Why do you think this whole business started? It's him. He's done it. He's brought the old man back and started the fear all over again.' She snorted impatiently.

'Who?' Jackus said quietly. He did not really need to ask. He knew what she would say. But she had to say it.

'*Who?* Marshall, of course.' And she stopped. Dead still. As if she had frightened herself by what she had said.

'Yes.' Jackus nodded, agreeing with the words she had not spoken. 'It's nonsense. You've been talking nonsense. You're just upset. Marshall is Marshall. Not anyone else. And the play's only acting.'

She looked stubbornly at him.

'Go on, Ann. Say it.' Suddenly it seemed desperately important that she should. 'Say "Marshall is Marshall".'

For a second longer, she was obstinately silent. Then she seemed to crumple. 'Marshall is Marshall,' she muttered, 'and I'm tired. I want to go home.'

Jackus pulled the door up so that it swung above their heads and watched her as she went down the drive like a sleep-walker. Then he picked up the garden broom and began to sweep the garage clear.

'. . . he didn't understand. He didn't understand at *all*. It was awful. They were so real to me. I felt the garage had been full of hundreds of them, pulling at my clothes with their little hands, crying and shaking. And all he was worried about was bits of bread and old lamp chops. It would have been funny if it wasn't so gruesome. Poor old Jackus.

All the same, he was quite nice to me. He didn't laugh, even though I cried all over the place and made an idiot of myself. And I don't think he'll tell anyone else about it. (It would be terrible if he did. They'd all be laughing at school and – I'd *die*.)

I wish I hadn't said what I did about Marshall. I forgot they were supposed to be friends. And I could see he didn't believe a word of it. But I'm sure I'm right. Marshall, when he's being Sweeney at least, has got something to do with all this.

Perhaps that's why he's being so horrible. If revolting Mrs Lovett and all those poor children could get inside my head, why couldn't that foul old man have got into Marshall's head? It's not so bad if I think of it like that.

If only there were something I could do to help them. I must, I must. No one's ever cared for them, and now I'm the only one who understands and wants to make things better. I must watch and wait, and perhaps if I do, I shall be able to see a way of changing things.

But I mustn't let Jackus know. He'll never understand. Never in a million years.

I must be very, very careful. . . .'

Chapter 11

Jackus knew what he ought to do next. But he put it off for as long as he could, trying to convince himself that it was unnecessary. Finally, on Sunday evening, he realized that the time had come. Pushing away his Maths homework, he stood up.

'I'm going out, Mum. Just for a bit.'

'It's rather late.' She frowned at the darkness outside. 'Where on earth are you going?'

'Thought I'd drop round and have a word with Marshall.'

No problem there. She beamed in approval and let him go without any more protests. Trudging through the orange-lit streets, he rehearsed in his head what he was going to say, setting his facts in order so that they ran as plausibly as possible.

But it still sounded fairly crazy.

Half-way to school, he turned and slid up a dark entry between two rows of back gardens. The sky was completely black and only a single old-fashioned lamppost shed a pool of light at the bend in the alley. Under his feet, he heard dead leaves shuffle and crunch. He crossed his fingers in his pockets, hoping that he had guessed right about where Marshall would be, that he would not have to go round to the front door and face Aunt Rose's suffocating, inquisitive welcome.

Just before he reached the lamp, he paused by a battered stretch of fence, listening. From the other side of the wooden planks he heard faint noises and, nodding in satisfaction, he slid one of the planks sideways and slipped through the gap into the bushes at the bottom of the garden.

Close to him, he heard someone move and the bright light of a torch beam was shone into his eyes, making everything else invisible.

'Jacko?' Marshall's voice laughed softly. 'Well, well. It's a long time since you've been here.'

'Can't you shine that torch somewhere else? It's scorching my eyeballs.'

The beam of light wobbled and then steadied as Marshall put the torch down on top of the cage in front of him. It lit the small area of garden in which they were standing, and illuminated Marshall's enquiring smile. From between his hands came a quick, murderous snap of teeth and his grip tightened round the ferret he was holding, his fingers scratching soothingly at its yellow head.

'What brings you here, then? After all this time? Been missing the ferrets, have you?'

'I want to talk to you.'

One dark eyebrow rose. 'I hope you haven't come to make a fuss. Fancy's a trifle nervous today. Aren't you, boy?' His fingers were gentle on the silky neck, but his eyes were just as watchful as the ferret's little red ones.

'No,' Jackus said quietly, 'I haven't come to make a fuss. I've come to explain something. About ghosts.'

'Really?' Marshall looked amused. 'I would have thought you'd had enough trouble with them already.'

'What?' For a moment Jackus was puzzled. Then he grinned. 'Oh, those silly ghost jokes. That was daft. Trust the others to make a game out of it. But I thought you might be more sensible.'

Fleetingly, Marshall smiled, as if at some private joke, but Jackus was too busy considering his next move to wonder why. 'Did you ask yourself,' he said carefully, 'why I stuck my neck out and talked about ghosts in the first place? When I should have guessed everyone would find it so hilarious?'

Marshall's finger went on scratching at the ferret's neck. 'I imagine that you were looking for an explanation for something that baffled you. That silly affair of Ann's sandwiches, perhaps?'

Jackus nodded. 'There was that. And Miss Lampeter's pencil flying through the air. And – one or two other things.'

'Like?'

'When I was sent down to the library that day. To learn my lines. I heard voices whispering and the books started to fall out of the shelves, all by themselves.'

'Dear, dear, that must have been very distressing for you,' Marshall said ironically. 'Especially as you cannot possibly expect me to believe you. If you have not got a witness.'

Unperturbed, Jackus went on. 'Then there was that day in the Art room. When we were painting scenery and everyone started to fight. Did it strike you there was anything odd about it?'

Marshall's eyes narrowed. 'A touch of hysteria, perhaps. People getting carried away for no reason. But plays are like that. Everyone gets screwed up tighter and tighter. It's very good for the performance. And the Lamppost was encouraging them to get at me. I mean –' he corrected himself with grave exactness '– at Sweeney.'

'And everything went wild,' Jackus said. 'Even the paint tins. They started falling over when no one was near them.' Marshall's mouth twitched, and he added triumphantly, 'You noticed, did you?'

'I noticed?' Marshall laughed. 'There was nothing to notice. You imagined it.'

'No I didn't.' Jackus was quite calm. 'I saw it. And when I stayed behind –'

'Oh yes, I wondered about that,' murmured Marshall.

'– I took these prints off the floor.' Reaching in his pocket, Jackus took out his crumpled piece of paper and handed it over.

'Here, hold Fancy,' Marshall said. 'You can give him some meat if you like.'

Jackus gripped the ferret behind the head and crooned to it, calming it with the sound of his voice. Then he reached over to a bowl on top of the cage, fishing out a lump of raw meat, faintly disgusted at its wet coldness. He dropped it into Fancy's mouth and the sharp little teeth chewed vigorously so that the juice ran out.

Marshall had unfolded the piece of paper and his head was bent over it thoughtfully. Jackus waited. He knew Marshall. There would not be a word until he had decided what his opinion was going to be.

At last he looked up and, almost absently, picked a piece of meat out of the bowl and fed it to the ferret with delicate fingers.

'I know what you're going to say,' Jackus put in quickly. 'You'll say I faked it. Did the prints myself with my own feet. But I didn't.'

'No,' Marshall said slowly. 'No, you didn't. You couldn't have done it with your feet. In fact, I don't suppose anyone in the school could have made marks like that.' For the first time he was totally solemn, his eyebrows drawn together and a long, dark crease running between them. 'They're most extraordinary.'

'Why?' Jackus leaned forward to look at them. 'What's so peculiar about them?'

'See how the toes spread out? Here, for example. And here. People who wear shoes have their toes all squashed up together, even when their feet are bare. These prints were made by people –children probably – who've hardly ever worn shoes.'

He went on staring, saying nothing, and after a moment Jackus grew impatient. 'Well?'

'It's very curious. There weren't any people like that in the room, and yet here are the prints.' He looked sharply at Jackus. 'Assuming you're telling the truth, of course.'

'Oh yes, I'm telling the truth.' Jackus did not need to protest. He knew that Marshall believed him. 'And everything you've said fits in perfectly. With what happened yesterday.'

Without speaking, Marshall held out his hands for the ferret. When he had it firmly in his grasp, he looked up. 'Come on, then. Let's have the next gripping instalment. What did happen yesterday?'

'Ann and I did an experiment. To try and raise the ghosts on purpose.' Jackus paused for a moment. This was the hardest bit to explain, because he did not fully understand himself what had happened. 'And we got what I expected. Footprints everywhere. Food chewed. Things taken. But there was something else.' He gulped, remembering it. 'Ann went really peculiar. She cried and moaned. She said she'd been feeling what the ghosts were feeling.'

'And?' The ferret slid smoothly between Marshall's hands, a streak of yellowish-white, flowing like water.

'And she said they were hungry, scared children. Like the Lamppost made us pretend to be, that day she had her stupid workshop. She said they were scared of a horrible old man who had them in his power. An old man like –'

'Like?'

'Like Sweeney Todd.'

There was a long silence. Marshall picked up a piece of meat and held it between his fingers, above Fancy's eagerly snapping teeth. Dribbles of blood ran down his pale skin. Finally, he said, 'Why are you telling me all this?'

It was not a question Jackus had expected. He gazed at the dark, shadowy bushes, trying to work out his answer. At last he said slowly, 'There's two reasons, really. The first is that I think you might be in danger. Because what the ghosts will want to do is get back at that old man.'

Startlingly, Marshall threw back his head and laughed, so suddenly that the ferret jerked with the shock. 'You mean,' he said, with every appearance of amusement, 'that I may wake up one night and find ghostly hands round my neck? Oh really, Jacko.'

'No, not like that.' Jackus's earnestness made him impatient. 'That wouldn't fit in at all. But they're very desperate. You should have seen Ann. And she said it was all directed at you.'

'I see. So little old Annie spins you a scary story, and you jump on your white horse to ride over and save me. Proper Jack the Giant-killer, aren't you?'

'Don't believe it, then,' Jackus said angrily. 'It's more comfortable to decide I've made it all up, is it?'

'No.' Marshall pursed his lips. 'I don't think you're clever enough to make it all up. And if you had, you'd be playing it more dramatic. Trying to pretend you were scared.' He suddenly grinned, almost friendly. 'But you're trying to pretend you're *not* scared, aren't you?'

Jackus shrugged. 'Dunno. But I know I don't like it.'

'And what's the other reason?'

'Eh?'

'You said you had two reasons for telling me. You've only given one so far.'

Looking him straight in the eye, Jackus said steadily, 'I think we ought to stop the play.'

Marshall tilted his head slightly sideways. 'You seem to have done that already.' Jackus just looked at him and he added sarcastically, 'I know. Don't tell me. It was the ghosts.'

'Yes,' said Jackus. 'That's what I think. And what I want you to tell me is what we're going to do about it.'

'*We?*' Marshall's face tightened. '*We* don't do things together. Not after last time. I like my allies reliable.'

'For Heaven's sake!' Jackus banged his fist hard down on top of the cage, startling the ferret so much that it nearly jumped out of Marshall's arms. 'Aren't I reliable enough for you? After all I've been through without telling? Why don't you *trust* me?'

'Because,' Marshall said, soothing Fancy, 'I don't trust anyone. It's like putting your head into a lion's mouth.'

'But people do do that. Sometimes.'

'Yes, and I'll bet there are an awful lot of headless people walking round because of it. I prefer to play safe.'

'But you can't with me, can you?' Jackus said. It came to him suddenly. 'You've got to trust me now, whether you like it or not.'

'So I have.' Marshall laughed lightly. 'Perhaps that's why the sight of you makes me so nervous.'

'Then you're an idiot. You ought to know I won't give you away. Ever. So why won't you help me, now I've asked you? Because you believe all those things I've told you. Don't you?'

Marshall's fingers moved smoothly through the ferret's fur while his eyes watched Jackus, weighing him up. Suddenly his clever, wary face relaxed into a smile.

'Why, of course. Of course I believe you, Jacko.'

'And you'll help me work out what we ought to do next?'

'I'll think about it,' Marshall said easily. 'I'll certainly think about it.' He was grinning now, gazing at Jackus with what seemed like openness.

Jackus felt himself relax with relief. It was as though

he had been holding his breath for weeks and now, for the first time was able to let it go. If Marshall was really prepared to come in on the problem, to think about it, then perhaps something could be done. And at least he was not alone any more, taking responsibility for Ann's screams and the wild, disjointed actions of the ghosts.

He grinned back and stuck out his hand. 'I knew you'd be O.K. if I really talked to you about it. It's ages since we've talked, isn't it?'

'Eons,' Marshall said politely.

'Well, it's good to have you on my side, anyway. Shake on it?'

Confidently, he stuck out his hand.

With a quick jerk of his head, Fancy jumped in Marshall's arms and turned, sinking white teeth deep into Jackus's finger. Jackus winced, letting out an involuntary yell. Slapping the ferret lightly on the head to make it release its hold, Marshall slid it back into the cage and shut the door. Then he looked down at Jackus's finger. Blood was starting to well, in slow drops, from the deep puncture marks.

'You ought to go home and put something on that. Before it turns septic.' He stepped backwards and pulled the loose plank aside. 'Don't want you landing up in hospital, do we?'

He ushered Jackus solicitously through the hole, murmuring, 'Don't forget.'

Walking off down the alley, Jackus sucked at his bleeding finger and glanced back over his shoulder. He could see Marshall still standing by the gap in the fence, a tall, motionless figure, silhouetted against the light from the lamp behind him. Jackus raised an arm, waving goodbye, but there was no response. Marshall just stood there, watching.

It had all been done so deftly, so quickly, that Jackus was half-way home before he realized that Marshall had not taken his offered hand.

Chapter 12

It was even later – Monday morning at least – when he realized that he did not know what his next step should be. Somehow, he had counted on some more active response from Marshall. Either complete involvement or utter disbelief. As things were, he was waiting for some sign.

All day, he waited. But Marshall was distant, not even smiling at him across the room. Finally, Jackus decided to make the first move. They were sitting in their class-room, supposedly beginning their homework, because it was too wet for football. Some people – the good ones – were writing busily. Even Ann, who had been quiet and aloof all day, was wrestling with her Maths. But Jackus was too confused to work. He kept glancing across at Marshall. And Marshall was abstracted, his fingers playing with something behind the cover of his book.

Suddenly, Jackus guessed what the something was. He got up and lounged across the class-room.

'Thought you were supposed to leave that in the props. box. The Lamppost'll have a fit if she sees you with it.'

Deliberately, Marshall opened the razor's gleaming blade.

'Guarding it, aren't I? The rate things are disappearing, nothing's safe unless it's in your hand.'

'Surprised you let me get within seeing distance, then,' Jackus said sourly.

He expected a clever retort but, instead, Marshall smiled up at him, a friendly, conspiratorial smile.

'Actually, I was just coming across to have a word with you. Why don't we push off, instead of sitting here like dummies for another twenty minutes? We could go up to the Hall and have a game of cards. If anyone sees us, we'll say we're waiting for the rehearsal to start.'

Jackus looked at him carefully. 'You mean – we could talk?'

'If you like.' Marshall shrugged. 'How about it?'

'I don't mind.' Jackus led the way out of the room and down the corridor. When they got to the Hall, Marshall began to pull the curtains across the tall windows, shutting out the grim half-light.

'Don't want anyone snooping through the windows at us, do we? And the Lamppost'll probably want them drawn when she comes. She might as well think we're virtuous and helpful.'

Jackus tugged at the curtain nearest him and ventured a question. 'Have you been thinking? About what I said to you yesterday?'

'Mmm?' Marshall glanced over his shoulder, looking vague. 'I –' Suddenly a flicker of irritation crossed his face. 'Oh, blast it. I've left my bag behind. Here.' He pulled a greasy old pack of cards out of his pocket and tossed it up on to the stage. 'I'll just nip back and get it. You start shuffling. Won't be a minute.'

As he ran lightly away down the aisle, Jackus vaulted on to the stage and sat on the edge, swinging his legs as he ran the cards backwards and forwards between his hands. He was feeling more cheerful than he had for weeks. If Marshall was prepared to be friendly at school, things might improve. Ever since the silly business of the ghost jokes, the others had all been hostile, eyeing him strangely and refusing to speak. And even Ann seemed to have gone funny now. She had been watching him oddly all day, but she had not made any attempt to talk. Blast them all.

'I'd like to cut their throats!' he murmured morosely. The words sounded pleasantly vicious in the empty Hall and, without really thinking, he added to them, adapting one of Marshall's lines. 'I'll cut 'em! From ear to ear. So the blood drips down their shirts and puddles on the floor.'

Abruptly, the lights went out. With the curtains drawn, the Hall was completely dark. Jackus jumped to his feet.

'Who's there?'

There was no sound.

'Oh, come off it. Put the lights on.'

As if in answer to his shout, there was a single soft, high note, clear as a flute. Right at the back of the Hall, under the balcony, a small pool of light appeared. In the centre of it lay something white, huddled in an indistinct mass. It twitched, lay still and then twitched again.

'Hallo?' he said, in a lower voice.

The white heap seemed to draw together. It rose slightly, making a low pyramid.

'Who is that?'

The pyramid swayed gently from side to side and rose another foot. It looked almost fluid, as though it had no firm substance, except at the top, where there was a more solid, defined round shape. Jackus whistled quietly through his teeth and the whiteness drew up and up, until it was nearly as tall as a man. The round shape at the top grew plainer. It had features like a face, but somehow not distinct enough for a real face.

111

For a moment, Jackus stood quite still, guessing. There was a chilliness round his ankles and he swallowed hard. At the back of his mind, he had been expecting something like this, sooner or later. But not quite like this. There was a feeling of wrongness about it, as though it did not quite tally with all the other oddities.

'Ha, ha!' he shouted defiantly. 'Think you're very clever, don't you? Well, I'm coming down to have a closer look at you.'

But he hesitated before jumping off the stage. In that second, the white shape drew up even taller, one foot, two feet, until it was impossibly tall, the face suspended eight or nine feet above the floor. The whole white pillar in its pool of light began to sway backwards and forwards, bonelessly, as no body could have done.

It was impossible to move now. Hardly possible to draw the heavy breath that came gulping up from his chest. His eyes strained in his head as he watched the insubstantial swaying, as the white shape rose and dipped and drew up tall again. He could feel a pulse thudding in his neck.

Then, close to his ear, just beside him, came the familiar half-audible whispering. And on his arm he felt a faint, small pressure, as though someone had laid a hand on him. The pressure increased, as though fingers were squeezing. Automatically, he chopped down with his other hand at where the arm must be. His fist passed through unresisting air, even while he could still feel the fingers squeezing.

He screamed, unable to stop himself.

At once, the light at the back of the Hall went out. Before he could move in the darkness, there was a sound of running feet outside. The next moment, switches clicked and he was standing blinking in the full glare of the lights, watching people hurrying up the Hall towards him.

They were all there. They must have been just arriving for the rehearsal when they heard his scream. Miss

Lampeter had her usual pile of books clutched to her chest, Benny and Stephen were whispering to some of the others, and Mandy was staring at him with frank curiosity. At the back, he could see Ann, her face white, as though she guessed what had happened. And behind her was the tall, stooping figure of Old Garner, poking his head round the door like a startled tortoise. Was he going to have to explain himself to the whole lot of them?

'Colin! Was it you that screamed?' Dropping her books on to the floor, Miss Lampeter vaulted on to the stage like a gymnast. 'What happened?'

'Yeah. Go on, Jackus. What was it?' A babble of voices sounded all at once, filling his ears with confusing questions. And Ann's eyes stared and stared.

Then, through the jabber, came Marshall's drawl. 'What's up, Little Jack Horner? Frightened of the dark, were you?' He came lounging up from the back of the Hall, with a smile on his face.

There were some giggles, but Miss Lampeter silenced them with a flick of her hand. 'Don't be stupid. Can't you see he's had a shock? Try to tell us about it, Colin.'

Jackus felt inexpressibly weary. First there had been the fright, and now there were all the questions and Marshall's mockery. He had no energy left to invent a plausible lie.

'I think I saw a ghost,' he said flatly.

That stopped them dead. People glanced sideways at each other and no one spoke. Only, at the back, Jackus saw Old Garner step quickly into the Hall and move up between the chairs.

'Oh yeah?' Benny said at last, in a voice that shook slightly. 'Didn't know you still believed in ghosts, Jackus. Only idiots believe in ghosts, don't they, Miss?'

Miss Lampeter was still looking at Jackus's pale face. 'The idiots are the ones who don't have open minds,' she snapped. 'What was it, Colin? Where did you see it?'

'Over the back there.' Jackus waved a hand. 'Under the balcony.'

113

As they glanced over their shoulders, Marshall spoke. 'Perhaps we should all go up on the balcony. Take a look.' It was dropped in lightly, even scornfully, as if it were a joke. But the others picked it up. As he might have guessed they would.

'Yes. Come on, Miss. Let's see what we can find.'

They had been waiting for some suggestion, something they could do. Now they started off down the Hall. Even Miss Lampeter slid off the stage to join them and Benny was half-way to the back when a quiet voice said, 'Stop. Stop there, Harris.'

With automatic obedience, they froze. Old Garner stalked to the front.

'I see no need,' he said acidly, 'for all of you to waste your energy charging up and down staircases. I'm sure Miss Lampeter is anxious to start her rehearsal.'

She gave a sulky half-nod and Mr Garner inclined his head.

'Exactly. Sit down, all of you, and wait for her to tell you what to do. Jackus,' he looked up at the stage, beckoning with one bony finger, 'you can come with me.'

As Jackus jumped off the stage, the headmaster gripped his shoulder and marched him away down the Hall. They had nearly reached the door before the chattering broke out behind them, almost drowning the sound of Miss Lampeter's voice giving orders.

'Right, boy,' Old Garner said. 'Let's go and look up on the balcony.' Pushing open the double doors at the back of the Hall, he led the way out into the corridor and up the narrow, dark staircase that curved round on itself. At the top, he paused and looked over his shoulder, smiling oddly. 'Shall we see what's on the other side of the door, then?'

It creaked as he opened it and Jackus gulped, watching it swing back into the darkness behind. But Old Garner's fingers were already on the switch and as it clicked the familiar prosaic clutter of the balcony came clear under the harsh light. The box for working the stage lighting. The heaps of old hymn-books and chairs. The tattered curtain that shut off the view down into the Hall. Jackus looked about, almost ashamed of himself for feeling so relieved.

'Nothing,' he said casually. 'Well, there wouldn't be, would there?'

'No spooks, certainly.' Old Garner sounded contemptuous. 'But I think we may find something to interest us.' He was darting his head around, the dry folds of skin on his neck stretching and slackening. Suddenly he fell on one knee and rummaged about under a chair. 'I think – yes, here we are.' He dragged out a bundle which had been thrust tightly into the space under the seat. 'I seem to have guessed right.'

Getting up, he dusted off the knees of his trousers and shook the bundle, letting the contents fall at Jackus's feet without saying anything. A Guy Fawkes mask. A tin whistle. A pocket torch with a wide cardboard collar, made to stop the light shining backwards. Or upwards. And fold upon fold of translucent white butter muslin, spilling towards the floor from the long black thread which was attached to one edge.

Jackus stared at them grimly. 'I've made an idiot of myself. Haven't I?'

'I don't think you need feel too foolish. Better men than you have been duped by methods just as crude as this.' Old Garner draped the muslin back over the mask and switched on the torch. As it wavered in his hand, the cheap exaggerated features momentarily took on eerie life. 'A very well thought out plan. I have no doubt you were badly frightened.'

'But why?' Jackus said, still not understanding. 'When I find out who did it I'll – kill him!'

'I see.' Mr Garner leaned back expressionlessly against the edge of the balcony. 'I should imagine it would give him great pleasure if you were to try. It would prove that he scared you. And it would force you to explain your gullibility to the others. He is probably counting on that.'

'He?'

'Oh, come on, boy. You must have some brains. Look down there.' With an impatient movement, the headmaster tweaked at the curtain behind him and jerked a finger over his shoulder. 'We'll name no names. But how many of them would have been clever enough to set you up so neatly? Eh?'

Jackus looked down. There on the stage were Benny and Marshall. Benny was gazing up, his fair, deceptively innocent face full of convincing fear and bewilderment. And Marshall was looking down at him. Confident. In control. Totally absorbed, for the moment, in dominating his terrified apprentice. 'If you breathe a word about what goes on in my shop, Tobias Ragg,' he hissed, 'I'll cut your throat from ear to ear.'

Marshall.

Jackus felt sick. He put out a hand to steady himself against the pile of chairs and looked from the figures on the stage to the heap of butter muslin and back again.

All the time, Old Garner was watching him, not unsympathetically. When he spoke, his voice was brisk and incurious. He dropped the curtain.

'Now I shall tell you how to deal with the situation. This is what you will do. You will say nothing. If anyone asks you what you found on the balcony, you can say that there was no one there. If anyone asks you about ghosts, it would be best to be non-committal. Simply go back to the rehearsal as though nothing had happened.'

'I'm not going back there,' Jackus muttered rebelliously. 'They didn't want me anyway, before this happened, and this puts the lid on it.'

Old Garner rubbed a hand thoughtfully across his forehead. 'I think,' he said slowly, 'that that might almost be part of the plan. Perhaps even its prime reason. Perhaps this person – whose name we are not mentioning – would be only too happy to see you out of the play.'

Jackus was startled. 'How did you –'

'Just a guess. A practised guess.' Old Garner shrugged. 'I have been making guesses about boys for more years than I care to remember. Will you accept my opinion and do as I ask?'

Grudgingly Jackus nodded and the headmaster gave a grim smile as he bundled the heap of oddments up in the length of muslin. 'I'll take care of these, then. I'm sure my wife can put six yards of butter muslin to some good use.' His knees clicked as he stood up straight. 'Let's go down.'

As they reached the bottom of the staircase, Jackus said, rather embarrassed, 'Sir.'

'Yes?' The beaky face looked round at him.

'It was lucky for me you came in, wasn't it? Otherwise they'd all have come charging up here. Found that stuff. They were all set to do it, before you stopped them.'

'I think that was the idea,' Old Garner said gravely. 'I don't *know*, but I would say it was another practised guess.'

'*She'd* have let them,' Jackus said, rather bitterly. 'She'd have led the way.' He grinned.

But Old Garner did not grin back. 'Don't be impertinent,' he said coldly.

'I didn't mean anything. I only meant I was glad you were there.'

'I'm sure Miss Lampeter is perfectly capable of taking charge of any situation that may arise,' Mr Garner said sourly. He turned to leave and then turned back, staring hard. 'Jackus?'

'Sir?'

'It occurs to me that this is the second time recently that you and I have been talking about ghosts. I am not a great believer in coincidence.' He cleared his throat delicately. 'Would you like to give me some explanation?'

Jackus's immediate thought was that he would never risk explaining anything to anyone, ever again. 'No, sir.'

'You are not, by any chance, being made a victim of a series of practical jokes.'

'No, sir. At least, I don't think so.'

'Very well.' The headmaster nodded. 'You are, of course entitled to your privacy. As long as what you do does not interfere with the proper running of my school. But if it does,' he paused, looking severe, 'I shall demand an explanation and I shall not be put off. Is that clear?'

'Yes, sir. Thank you, sir.'

Old Garner nodded again and stumped off, the untidy bundle of butter muslin clutched under his arm. As Jackus turned to go back into the Hall, he found himself shaking. But not with fear. With rage. He kept remembering Marshall's friendly smile of the night before. That lying smile. It was all he could do to hold himself steady as he pushed the door open.

When he went in, everything stopped. Benny jerked his arm away from Marshall's hand and yelled, 'Find anything, Jackus?'

Instead of scolding him, Miss Lampeter looked round inquisitively. But Jackus did not even glance at her. He walked up the Hall without meeting anyone's eyes. Especially not meeting Marshall's.

'What happened, esteemed and ridiculous Jackus?' Stephen said. 'The burningfulness of the curiosity down here is indeed terrific.'

'Nothing happened,' Jackus said in a matter of fact voice. 'There wasn't anyone there.'

'Ah – so you think it was a *real* ghost?' Marshall muttered sarcastically.

Jackus had no intention of falling into that trap. 'I'm keeping an open mind,' he said gravely. 'That's the only thing to do, isn't it, Miss Lampeter?'

'Yes, Colin. Very wise.'

He hardly heard her. He was too busy noticing the

quick, almost imperceptible flicker of Marshall's eyes as he looked away. The others were disappointed. They went on with the rehearsal woodenly, cheated of their bit of excitement.

Ten points to me, Jackus thought. But he was far too angry to feel smug. He sat rigidly in his chair, trying to keep his face expressionless. The struggle kept him so busy that it was some time before he remembered the fingers that had gripped his arm in the dark.

How on earth had Marshall managed that?

As he thought it, simultaneously, he felt fingers on his arm again. He glanced round, startled. But these fingers were all too solid and real.

'Please,' whispered Ann, 'I can't bear it. You've got to tell me what happened.'

Monday, 7th December

'... and at first I thought I wasn't going to get him to talk at all. He was like a block of wood. Just sat there, with his lips pressed together. In the end, I followed him home, because I *had* to know. And when we got to the canal, while we were walking along the towpath, he gave in.

I didn't understand in the beginning. All he would say was "Marshall. It was Marshall." Then, when I got what he meant, I was furious. To think he's actually told Marshall! Warned him! Now he'll be on his guard, and that will make things harder for Them. As if everything wasn't hard enough for Them already, poor little things.

There was just one moment, one lovely moment, when I thought it might all have been Marshall. All the whole peculiar lot of it. But that didn't last long. I only had to think to see it wouldn't be possible. Anyway, I don't believe I could have imagined Them, however mad I am.

The only good thing is that Jackus has stopped all that rubbish about being friends with Marshall. His face was quite white, and he kept saying, "Why did he do it? Why did he suddenly turn against me? When he was so friendly on Sunday?" So I told him. All about how it was Marshall's idea to tell those jokes about ghosts and get him into trouble. And he wasn't white any more after that. His face went red, as if he was very, very angry.

I nearly told him the rest, then. I think I could have got him to understand that people like Marshall can't be allowed to go on taking and threatening and using. Like an awful spider. No, that's not bad enough. Like a horrible fungus, growing and growing and making everything ugly and unbearable and –

Well, anyway, he wouldn't listen to me. Jackus, I mean. When I started to talk, he just pushed at me, so hard that if I'd been little like Mandy I'd have landed in the canal. And he said, "Leave me alone. Just leave me alone." In a funny voice, as if he was choking. Then he rushed off.

And it's ALL MARSHALL'S FAULT. Poor old Jackus, I felt quite sorry for him. But feeling sorry for people isn't enough. It's feeble by itself.'

Chapter 13

It was no better when he opened his eyes the next morning. The hard truth remained: Marshall had played a cruel trick on him, a trick designed to demonstrate that they were not friends or allies. And instead of shutting it close inside himself, he had been persuaded to blurt the whole thing out to Ann, which had only made it worse.

Jackus groaned and rolled over, burying his face in the pillow. He did not see how he could ever go to school again, and certainly not this morning. He would have to convince his mother that he was ill.

By the time he went down to breakfast, he had more or less convinced himself that he had a terrible stomach-ache. He picked at his cereal without enthusiasm and was just about to tell her how awful he felt when she looked up brightly.

'Oh, I meant to tell you. Rose is coming to tea this afternoon. Wouldn't it be nice if you brought Big Colin home with you?'

'Can't,' he said quickly. 'We've got a rehearsal. Very important. Last one before the dress rehearsal. I'll probably be home quite late.'

Well, that was it. He had committed himself to going to school now. But anything was better than a cosy tea-party, with their mothers reminiscing about what good friends they'd been when they were little boys.

The trouble was that the stomach-ache was very real by now. It might only be caused by nerves, knotting his muscles together, but it was giving him genuine pain. He tottered down the road, only just getting to school before the bell, and spent most of the day hunched over his desk, unable to concentrate on anything.

From time to time, people made unusual efforts to talk to him. In the middle of Maths, Benny leaned over and whispered, 'Thought any more about that ghost, Jackus?'

He shook his head grumpily, not saying anything. It was only when first Mandy and then Stephen asked the same question that he realized that Marshall had done him a favour, in a queer, back-handed sort of way. The others, who had been so hostile before, were intrigued now and ready to let him be one of them. He could have been part of one of the chattering groups that gathered at the end of each lesson. If only he had not felt so rotten.

But he did feel rotten. What had happened yesterday went round and round in his head, keeping him glum and silent and worsening the miserable ache that seemed to spread all over his body now. He was so wrapped up in it that he did not think about the play at all, and he did not remember the significance of that afternoon's rehearsal until, when they were on their way down to the Hall, he heard Stephen mutter to Benny, 'Well, this is it. Unless something happens, it'll be the last time we do the play.'

'What do you mean?' Jackus interrupted, his attention caught.

They both glared at him.

'That's typical of you, Jackus,' Benny growled. 'Fat lot *you* care about the play. I bet you'll be glad if those things aren't returned and Old Garner puts the shutters on it.'

'Oh, I'd forgotten all about that.' And he had. He was amazed at himself. If the play were stopped, that would end one set of problems, at least. Feeling slightly less gloomy, he followed the others into the Hall.

As soon as he saw Miss Lampeter, he knew that she, at least, had not forgotten. She was standing on the stage looking down at them all as they came in, and her whole appearance showed a kind of wildness, as if she were distraught. Her fair hair hung round her face in untidy hanks, her blouse had come untucked from her skirt and her shoes were unpolished. Instead of her usual brisk cheerfulness, she was full of grim determination, her face set tight. With an impatient hand, she motioned to them all to sit down in the chairs at the front of the Hall, and when everyone was there, she started to speak.

'I need not remind you, I'm sure,' she said coldly, 'that one person's selfishness is about to throw away all the work we've done so far. I had hoped that all the stolen things would be returned by today, so that we could concentrate on putting on a decent production, without all the suspicion that has got mixed up with it. Instead,' she gulped, 'instead things have got *worse*.' She gave an apologetic glance at Marshall. 'I'm sorry, but someone has taken your razor out of the props. box.'

'Oh, that.' Marshall laughed. It was an odd sound in the tense silence. 'No, that's all right, Miss Lampeter. It's my fault, I'm afraid. I was worried about it, so I thought the best thing was for me to take care of it myself.' He gave her his most charming smile.

But she was too overwrought to be calmed, even by Marshall's smiles. 'I told you,' she said angrily. 'I said

124

you were to leave it in the box. What do you mean by taking it out without my permission?'

'I'm sorry,' he said, not looking in the least perturbed.

'Well, give it back at once. Even if the play is going to be ruined, I will have obedience.'

'All right, all right.' Negligently, Marshall flung open the lid of his case, to take out the razor.

And stopped.

Mandy, who was sitting beside him, leaned over, mildly curious, and her face suddenly twisted with accusation.

'Marshall! That's my cross! And –' She broke off, staring at him, as though she could not believe her eyes.

Very slowly, Miss Lampeter walked down the steps from the stage and looked into Marshall's case. Then she straightened. Without glancing at Marshall, she said icily, 'Stephen, I think you should go and fetch Mr Garner.'

While they sat in silence, waiting, Jackus craned his head to try and see what Mandy had seen. There, ranged in a neat layer on top of Marshall's books, was a whole array of things. Mandy's cross and chain. Benny's calculator. A pound note. A handkerchief. A watch. Before Jackus had finished counting them, Old Garner was walking up the Hall with Stephen trotting along behind.

Old Garner did not say anything for a moment. He studied the contents of the case thoughtfully. Then he raised his eyes and looked directly at Marshall.

'Well? What have you got to say?'

Marshall shrugged. 'It's ridiculous, isn't it? If it was me that had taken those things, I wouldn't be so stupid as to carry them round in my case and open it for everyone to see.' He sounded cool, but his face was pale and Jackus could see a vein throbbing underneath his eye. He felt a quick, nasty twinge of pleasure. Marshall was in a hole, and he knew it.

'Everyone couldn't see,' Mandy said slowly. 'If I

hadn't happened to look, no one would have known at all.'

Everyone was staring now. Marshall looked proudly away from them all, at Old Garner.

'I think you have a case to answer,' the headmaster said gently. 'How do you propose to defend yourself?'

'They must have been planted, mustn't they?' Marshall sounded a fraction too defensive and Stephen gave a scornful snort of disbelief.

'D'you expect us to believe *that*?'

Then, abruptly, Marshall lost his temper. He did not leap to his feet or make grand gestures, but Jackus knew, because he went very quiet, very still, his lips drawn slightly back over his teeth.

'I can't think what all the fuss is about,' he said languidly, almost as if he were bored. 'It's obvious to anyone who knows that I'm not likely to be the culprit. Not when we've got a thief in the cast anyway.'

Jackus felt a twinge as his stomach knotted more tightly.

'What do you mean?' Benny said. 'Who's a thief?'

Don't say it, Marshall, Jackus thought. *You can't do that.* But Marshall did not even hesitate.

'Why, Jackus, of course. Didn't you know?' He leaned back in his chair and glanced round, amused, as everyone's eyes jumped to Jackus's face. 'Why do you suppose he's in the play at all? Mr Garner caught him in the school at night, at half-term. He was busy nicking a couple of cassette recorders. Isn't that right, sir?'

Jackus squirmed. The accusing eyes were on him now, and he knew that he ought to say something to protect himself. But he could not tear his gaze away from Marshall's smiling face. How could he have done it?

In the end, it was Old Garner who spoke. 'Yes, Marshall,' he murmured, 'I did find Jackus, in just the circumstances you describe. But how did you know about it?'

Marshall's smile broadened. 'Why, Jackus told me about it himself. Didn't you, Jacko?'

All at once, Jackus felt inexpressibly weary. He knew that he should have attacked in his turn, explained how things had really been. But what was the point? Everyone knew now, and nothing he could say would make any difference to that. It would just be playing Marshall's game. And he found, rather to his surprise, that he despised Marshall. So he nodded. Let them think he had stolen their stupid little bits and pieces. What did he care?

'You see?' Marshall murmured. 'I don't seem to be the obvious suspect, do I?'

Old Garner coughed, preparing to speak, and Jackus thought, *This is it*. But what the headmaster said was completely unexpected.

'I am not as blind as you seem to suppose, Marshall. Even headmasters have their moments of insight. You may not have realized, but I have already discussed this matter with Jackus, and he has convinced me that he had nothing to do with it. I do not see why I should change my opinion just because you have given away something he told you in confidence.'

Jackus felt his mouth drop open. So, Old Garner really had believed him. It had not been just a trick. The thought was strangely comforting.

But it obviously did not please Marshall. 'Dear, dear,' he said acidly. 'I suppose that leaves me as the criminal then.'

'No.' It was Ann. All the time everyone else had been speaking, Jackus had been faintly aware that she was twisting her hands together and growing pinker and pinker. She broke in now with such an effort that the words came out like a splutter. 'No, it wasn't Marshall. It can't have been. My watch is there.'

'Yes?' the headmaster said.

'Well – I didn't lose it at school at all. And Marshall wasn't there either. He can't have known where I was even.' She finished, panting, as though she were relieved that she had made herself say it.

'Ah, the plot thickens,' Mr Garner said ironically. 'Are there any more odd bits of evidence?'

'Does it matter?' They had almost forgotten about Miss Lampeter, up on the stage. She stood there, gripping her hands together like a little girl asking for permission for something. 'The things have been returned, even though we haven't caught the culprit. That means we can go on with the play, doesn't it?'

Mr Garner gave her a smile of pure affectionate amusement. 'Yes, it does, my dear. I promised it would be all right if the stolen property were given back.' Then he frowned. 'But the thief – whoever he is – need not think the matter will be dropped. I shall go on making investigations. But I shan't cancel the play.'

Visibly, Miss Lampeter brightened. Her clothes were still untidy, but she became brisk and enthusiastic once more. With a quick gesture, she tucked her blouse into her skirt and said crisply, 'I think we ought to get on with the rehearsal, then. We've lost half an hour already, and there are lots of loose ends to tie up before Thursday.'

'I'll leave you to it.' Gravely, Old Garner started to walk away as she began to hustle everyone up into the wings. Jackus felt slightly shell-shocked. He had been braced for some kind of catastrophe and now everything had returned abruptly to normal. But not quite to normal. As people got out of their chairs and moved towards the steps, they glanced at him with more than the usual hostility. It was clear that, however much Old Garner believed in him, they were not all convinced.

But Ann, as she passed him, leaned sideways to whisper in his ear. 'It's all right, Jackus. I know it wasn't you. It was – you know.' Her face took on a faintly vicious expression. 'But I wish it *had* been Marshall.'

Marshall himself was right behind her. He did not hear what she said, but as he passed he gave her a mock bow. 'I suppose I should thank you for saving my reputation. Most kind. But what a pity it was that you couldn't have done it sooner. It would have avoided a lot of – nastiness.'

Ann tossed her head. 'You're lucky I said it at all,' she muttered coldly. 'You deserve to get into trouble, Marshall. With all the rotten things you've done lately. It's just that I want to see you squirm for something you really did do, and not just for a misunderstanding.'

'What it is to have scruples.' Marshall smiled annoyingly. 'You'll never make a good enemy if you're so feeble and soft-hearted.'

She stared at him for a second, as if what he had said were much more important than the actual words. Then, tossing her head again, she marched off towards the stage, where Mandy was sitting wringing her hands, ready to begin a scene of pathetic lamentations for her lost lover.

Jackus and Marshall were left facing each other.

'You didn't need to do that,' Jackus said in a very low voice, almost choking. 'It was cruel.'

'It was self-defence.' Marshall glanced up quickly at the stage, to make sure that they were not observed, and then looked back at Jackus with a hard stare. 'You were trying to get your own back, weren't you? For my little joke, yesterday? Well, it didn't work, did it? As soon as I saw those things in my case, I knew you'd planted them there.'

'You're crazy. I wouldn't do a thing like that.' Jackus was so startled that he protested too loudly and Miss Lampeter gave him an irritated glare. Dropping his voice, he added. 'It wasn't me. It was part of what's been going on all the time. It was –'

'Your precious ghosts?' Marshall shook his head. 'Do you think I believe that? No, it was you. And you got your fingers burnt, didn't you? It's your own fault if everyone knows you're a thief now.'

'It would have served you right,' Jackus said hotly, 'if I'd told them what you had to do with that.'

'Ah, but I knew you wouldn't do that. Because you promised to keep it a secret. And you're like Annie. Too feeble and soft-hearted to make a good enemy.'

'You promised too,' Jackus hissed. 'And it didn't stop *you* telling.'

'No. But I have an advantage over you, you see.'

'You have?'

Marshall smiled charmingly, as if he were about to pay a compliment. 'Yes, I have. You see – I've never liked you much.'

As he walked past, towards the stage, he was actually chuckling.

Chapter 14

Jackus ran a finger round the tight waistband of his trousers and straightened the tattered rags which hung below his knees. Then he turned and looked into the mirror. It was a good costume. Desperately worn and patched, but jaunty in its crude, bright colours. It should have put him perfectly in the mood for being cheeky Jarvis Williams. Only he felt nothing except a detached coldness. The play was only words to him now. Somehow he had to get his way through the dress rehearsal and the three performances, among cold, silent stares. And he was not able to think any further ahead than that.

He pushed open the door of the domestic science room and walked down the stairs which led to the back of the stage. Almost everyone else was there already. Mandy, pretty and slender in a tight-waisted dress, was spinning round, setting her crinoline skirts swirling. Stephen was stroking the black moustache that had been glued to his upper lip, and Benny was wriggling awkwardly, trying to get comfortable inside the jacket he had been given, which was purposely too small.

Only Ann was completely still. She sat slumped on a chair, her hips padded out so that she looked even larger than usual. From time to time, people bumped into her, but she did not even glance up. Just went on staring at the ground, blank-faced.

Jackus walked across to her. The expression on her face exactly matched how he was feeling, and he felt she was the only person he could bear to speak to. Even if she had not been the only person who was likely to answer him.

'You look good,' he said politely. 'Just right.'

'Oh. Yes.' Her voice was dull. 'Horrible, you mean.' The make-up on her face was thick and pasty yellowish-grey and someone had drawn delicate lines round her eyes and the corners of her mouth.

'Cheer up.' He did not feel at all cheerful himself, but her gloom was suffocating. 'It's only a play. Remember?'

'How can you say that?' For the first time she roused herself, glaring up at him. 'You ought to know it's not true, even if everyone else is too thick to see it. I just wish it was all over.'

'Well, it will be soon. By Monday.' He glanced round at the others. They were all there now. Except one. He did not say anything, but Ann seemed to read his mind.

'No, he hasn't come yet. I expect he's waiting to make a grand entrance.' She looked suddenly venomous. 'I hope he falls down the stairs and breaks his rotten neck!'

'Want me to sneak up and give him a push?' Jackus said flippantly. But she did not smile.

'I wish you would. That would get rid of him and the beastly play, all in one go.'

Before Jackus could reply, there was a sudden stir. People turned to look towards the staircase, nudging each other and whispering. Even before Jackus spun round, he knew what he would see.

Marshall was coming slowly down the steps. His lips were curved in the familiar, mocking smile, but the rest of his appearance was eerily transformed. In the full shirt and the tight trousers, with the barber's apron round his waist, his body was wiry and lean, its stooping giving an impression not of feeble age but of malevolent energy coiled up and held in check. His hair had been powdered to a uniform grey colour and, beneath it, his eyes glittered with arresting brightness. In his right hand he held the razor, its blade sheathed in the polished wooden handle.

'Isn't he *lovely*?' Until she spoke, none of them had realized that Miss Lampeter was behind him. Now she clattered down the stairs, jumping the last two, and grinned round at them all. 'You're all lovely. I feel I've gone back a hundred years, standing here looking at you.'

'Marshall looks about fifty years older,' Stephen said gravely. 'I wouldn't like to meet him on a dark night.'

'Or at all,' murmured Jackus. But not loud enough for anyone except Ann to hear.

Miss Lampeter clapped her hands. 'Right. Let's get started. All go to your places. Is Mr Gregory there with the music?'

'Ready and waiting, boss.' Mr Gregory was a large man with a round red face who pounded the piano enthusiastically. 'Just give the word when you want me to start.'

A few moments later, he raised his hands and brought them down in a series of crashing chords which thundered about the Hall. They quietened gradually, dying away to a thin thread of sound, slow and creeping, and the curtain rose, revealing the outside of Sweeney Todd's shop. The whole stage was in darkness except for a pool of light in the centre.

'Marshall,' murmured Miss Lampeter under her

breath. But even before she had spoken, he was sidling out from behind the curtain at the side, towards the patch of light.

'They all come –' he began.

Immediately, everything went dark.

'Lights!' shouted Miss Lampeter. 'Oh, what are you doing up there?'

'Sorry,' came a muffled voice from the balcony. 'Don't know what happened.' Feet shuffled. There were flashes of red, green, yellow light. 'Think the bulb's gone. Hang on a minute.'

Miss Lampeter sat tapping her foot impatiently and, in the wings, Jackus felt someone stir beside him in the blackness. 'It's going to be terrible,' muttered a voice in his ear. 'Can't you tell? Everything's waiting and –'

'You're imagining it,' Jackus said gruffly. He was in no mood for dealing with a hysterical Ann. 'Just keep calm.'

But he found that the palms of his hands were sweating. He rubbed them on his ragged trousers and wished that he did not have to wait so long before his first scene. If only he had something to do, it would not be so bad.

Marshall was gliding through his opening speech now, apparently unruffled by the setback, and for a while things went smoothly. Whenever Jackus looked at Ann, he could see her chewing her fingers or screwing her hands together absently in a fold of her skirt, but he ignored her. She was just getting worked up about nothing.

Whatever the reason for her tenseness, it certainly improved her acting. Her first entrance came quite early on, when she tried to persuade Sweeney that they should give up their horrible trade. As she walked on to the stage, her face was set into an expression of genuine fear and she cowered away from Marshall as though she found him both disgusting and terrifying. Gradually, as she pleaded, Marshall edged her into a corner, hardly speaking, but dominating her with his eyes. The scene was supposed to end with them sitting side by side listening to the clock strike midnight, while Sweeney's fingers rested lightly but ominously on Mrs Lovett's arm. 'It's a demonstration of Sweeney's power,' Miss Lampeter had insisted at rehearsal after rehearsal. 'Quiet but strong. You two must build it up, so that by the time the clock is ready to strike the audience are sitting on the edge of their chairs.'

Now, for the first time, it seemed that they had got it perfectly. Ann did not need to make any dramatic gestures. Marshall did not need to bluster or rant. It was obvious to everyone watching that the two of them were engaged in a desperate silent struggle. Jackus found himself digging his nails into the palms of his hands and staring out at the stage almost without blinking as the clock began to strike one . . . two . . . three. . . .

Then it reached twelve and, at a single blow, the tension was shattered. In the wings and down in the Hall, people started to laugh uncontrollably. Because it simply went on. Thirteen . . . fourteen . . . fifteen. . . .

'George!' Miss Lampeter yelled furiously. 'What's going on?'

'Sorry,' said the voice from the balcony again. 'The record's stuck. I can't –'

There was a crash, magnified over the loudspeaker system, and then, mercifully, the striking stopped and the giggles began to die down.

'Go on, everyone,' Miss Lampeter said, grimly determined. 'You mustn't let yourselves be put off by things like that.'

Marshall stalked off the stage, walking past Jackus without even a glance. He went on moving in the same hunched, stiff way right to the far side when the audience could no longer see him. Sitting down on a chair, he gazed at his feet, oblivious of everyone else.

'Look at him!' Ann hissed as she came level with Jackus. 'He doesn't change, does he? Even when he's not acting. Sweeney's got right inside him.' She was visibly shaking.

'I expect he's just trying to keep in the mood for his part,' muttered Jackus, offhand. He did not want to think about Marshall. Ann looked at him scornfully.

'Still making excuses for him, are you? Can't you see it's not just a part?'

'Sssh!' Jackus turned away, to put her off, and watched Mandy glide on to the stage from the other side, wringing her hands woefully.

She was trying hard, but she seemed to have lost the spirit she had shown at rehearsals. She kept glancing about her nervously, as if she were waiting for another disaster. Even from where he was standing, Jackus could hear Miss Lampeter's impatient sighs. But nothing did happen. Not until Marshall's next entrance.

As he moved out of the wings, on to the lighted stage, Mr Gregory began to play the creeping music that always heralded his entrance. In dumb show, Marshall welcomed the jeweller who had come to be shaved and ushered him into the chair. While they talked, the music continued in the background, low and menacing, and when Marshall went out and passed behind the gauze, pausing with his razor uplifted, there was a shivering trill on the piano that exactly matched the threatening gesture.

What happened next had never been properly rehearsed, but it looked as though it was going to be very effective. As Marshall re-entered, Tom, the jeweller, held up his string of pearls to admire. Sweeney's eyes gleamed and he stepped forwards, brandishing his razor, as the piano crashed out a trembling chord.

The timing was perfect, the notes sounding at precisely the same moment as Marshall's face lit up with greed. But instantly there was a loud incongruous twang. Marshall's hand jumped, Tom gave a wild yell and Mr Gregory burst out laughing.

'God!' said Miss Lampeter ferociously. 'This *bloody* rehearsal. What is it *now*?'

'Amazing thing,' boomed Mr Gregory cheerfully. 'Never known it happen before. One of the piano strings has snapped.'

'Who cares about the string?' shouted Tom in an injured voice. 'Look at my hand!' He held it up. Across the knuckles was a thin red line, from which blood ran slowly down his fingers. 'Marshall cut me with his wretched razor.'

'Marshall!' Miss Lampeter bellowed. 'I told you to be careful with that thing.'

'I'm sorry,' Marshall said. 'The piano string made me jump.' He stared down at the reddened blade of the razor, as though realizing for the first time how sharp it was.

'But what about my *hand*?' whined Tom. 'I can't just leave it like this.'

'Don't be such a baby!' Miss Lampeter did not sound in the least sympathetic. 'It isn't serious, is it?'

'It's *bleeding*.' Tom was very tenacious. When he had a grievance, he did not give it up easily. Miss Lampeter blinked and looked closer.

'Oh yes, I'm sorry, Tom. It is nasty, isn't it? You'd better go to the medical room and get a plaster on it. We'll just have to cut the rest of your scene. And please concentrate on what you're doing, Marshall. Things are bad enough, without you playing the fool.'

She slumped down gloomily in her chair and waved a hand at the stage, to indicate that they should go on with the next scene. And the rehearsal limped on its way. But everyone was thoroughly rattled now. Even without any more mechanical disasters, things kept going wrong. People entered too early or too late, forgot their words and tripped over the scenery. Jackus was trying hard to keep calm, but he found the general air of distraction catching and as he charged on to the stage to rescue Benny from the madhouse he slid on a handkerchief that someone had dropped and had to clutch at Stephen to save himself from falling.

'Colin!' Miss Lampeter snapped. 'Don't be so clumsy!' But there was no real bite to her voice. She sounded as though she had more or less given up.

Only Marshall seemed completely absorbed in his part. Unaffected by everyone else's mood, he stalked on and off the stage, his timing perfect and his eyes growing more and more insane, as he built up Sweeney Todd's increasing frenzy, his sense that his enemies were closing round him. When he was not acting, he stood in the wings, watching the stumbling efforts of the others with a supercilious smile.

And they glared back at him.

Gradually, the play lurched its way towards the final scenes, in which Sweeney, escaped from prison and gone completely mad, returned to the cellar in which Mrs Lovett lay dead. Talking dementedly to her body, he hunted frantically round the vault for the jewels which had made him murder.

It was one of Marshall's best scenes. The madder he grew, the quieter his voice became, while he rummaged from table to chest and from the chest to the tall shelves which stood at the back of the stage. Now that there was no one else round him, to spoil the atmosphere, he dominated the Hall, an evil, powerful presence.

Jackus, standing in the wings ready for his final entrance, found himself admiring reluctantly. Marshall might be cold and cruel, but there was no doubt that he could act. And perhaps – the thought came as a kind of feeble consolation – perhaps Ann was right and the play had affected him, soured his temper and made him suspicious. When it was over, things might be better. Perhaps all the things he had said really meant nothing.

Jackus was thinking so hard that it was a moment before he realized what was happening. Marshall, having bent over Ann's supposedly dead body to croon a few words, straightened and moved backwards, to the shelves. As he approached, they trembled slightly and then began to tip. Jackus started forwards, sticking out a hand to catch them, but it was too late.

They crashed to the floor of the stage, narrowly missing Marshall and sending a heavy wooden thud resounding round the Hall.

For a second it was the only noise. Everyone was stunned, and Marshall turned white and put a hand to his face. The shelves had fallen close enough for him to feel the air move as they passed.

The next moment, Miss Lampeter was on the stage, yelling. She strode to the wings and dragged Jackus out, her fingers gripping his shoulder viciously.

'You stupid, wicked boy!' she was shouting. 'You could have killed him! You could have *killed* him! That was a frightful thing to do!'

All the time, she was shaking him hard, so that his head jerked chokingly backwards and forwards, forcing tears into his eyes. Her hair tumbled over her face and as her mouth twisted into accusations bubbles of spit appeared at the corners. And the shouting went on and on.

'You've been a menace all the time! Stealing, sneering, breaking things up! I should never have had you in the play. I should never have tried to help you. You're stupid and hopeless and *evil*!'

'I didn't!' Jackus gasped out. 'I wasn't doing it! I was trying to –'

Another fierce shaking left him speechless. He was just about to lash out at her, unable to bear it any longer, when he heard a voice say, 'It's all right, Miss Lampeter. Don't upset yourself.'

Old Garner had come quietly up the Hall, bringing Tom back with his fingers bandaged. Through the mist of blood that jumped and danced behind his eyes, Jackus could see the headmaster's arm firmly round Miss Lampeter's shoulders, apparently soothing her, but actually holding her back.

139

Then everything swam and he had just enough consciousness left to hear Old Garner say, sharply, 'Sit
down and put your head between your knees. Quickly,
boy!'

The headmaster's face blanked out, and the last thing
Jackus felt was the boards of the stage coming up to hit
him on the cheek.

Thursday, 10th December

'. . . and they couldn't *see* it. None of them. Not even Jackus. I know he's not like me. I know he can't feel when They're there. He doesn't suck up the fright and the anger, the way I do. And that was bad. Very bad. The anger was so strong I had to keep biting at my lip so that it didn't make me shout out.

But even Jackus ought to have seen that it was all pointed at Marshall. Everything that went wrong, all the disasters, happened when he was on the stage. They were meant to get him. To stop him going on and on and on being Sweeney in that horrible way. And the terrible thing, the thing that really upset me. . . .

I nearly couldn't write it, because it frightened me to make the words with my pen. But it's *true*. The terrible thing was that They couldn't do it. They're not strong enough by themselves. They can rage and fling things about and break things, but They can't get at him and hurt him like he ought to be hurt and stop the blackness and the fear that comes whenever he steps on to the stage and does those awful cruel things and speaks in that beastly voice as though he had everyone in a tight tight grip and would never let go ever but make the fear go on and on so that even when you're dead you don't realize, you don't see any way you can get free from that choking choking. . . .

They can't do it by themselves. Poor, poor little things.'

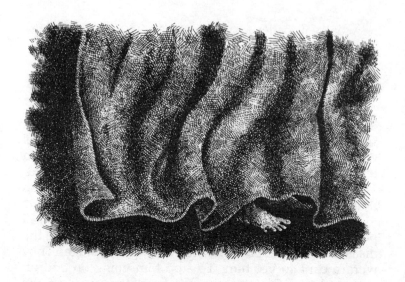

Chapter 15

'Enough is enough,' Old Garner said. He sat at his desk, his fingertips resting together in a neat arch, and looked straight across at Jackus, who was sitting on the other side. 'I told you once before that I would respect your privacy. But I also said that if whatever you were doing began to interfere with the running of the school, I should demand an explanation. Now the time has come for that explanation. What happened last night was not the kind of thing I can ignore. And, once again, you were at the centre of it. What have you got to say to me?'

Jackus sat for a second, considering. He had meant to come anyway, even if Old Garner had not sent for him as soon as he arrived at school. But now he was here, he was not quite sure how to start.

'I don't think you're going to believe me, sir.'

The headmaster tilted his head slightly to one side. 'Perhaps not. But truth has an authority of its own. Tell me what you believe to be the truth, and I will listen to you with an open mind.'

He did not say anything else. Just sat very still, waiting, his grey eyes cool and quizzical. Jackus cleared his throat.

'Well, this play we're doing, it's not true, is it? I mean, there never was a Sweeney Todd who killed people in a barber's shop and had them chopped up and made into pies?'

Mr Garner shook his head. 'Not as far as I know.'

'But in another way it is true, isn't it? Then – in Victorian times – there were lots of people who were very cruel to other people, to make money out of them. Especially to children. There must have been lots of children who spent their lives being afraid and hungry and angry at the world. Mustn't there?'

'Yes, I'm afraid there must.' Mr Garner looked as though he were about to add something, but he stopped himself. 'Go on.'

Jackus looked down at his fingers. 'And perhaps, if what they felt was strong enough, it might go on and on – kind of hanging in the air – so that it carried on after they were dead. Perhaps, even stopped them knowing they were dead?'

He had expected a laugh, or some scornful comment. But there was nothing. Only the same attentive gaze. He searched for words to explain the rest.

'And if someone, or a group of people, started to try and reproduce those feelings, like Miss Lampeter's tried to get us all to pretend to be terrified of Sweeney Todd, it might, well, it might strike a sort of echo. D'you think?'

He felt his voice die away lamely and he glanced up. Old Garner was staring at him with interest. 'Like the wineglass,' he said softly. Seeing Jackus's bewildered expression, he went on. 'I have always understood that if someone sings a high, powerful note, a wineglass in the room will vibrate in sympathy, to the same note. And sometimes –' He hesitated.

143

'Sometimes what?' Jackus said quietly.

'Sometimes the wineglass will shatter with the strain.'

Jackus stared back, digging his nails into the palms of his hands. 'Yes,' he murmured. 'Yes, that's just what I mean.'

Mr Garner rested his chin on his arched fingertips. 'I think you should tell me what evidence you have for saying all this. You have got some evidence?'

'Oh yes.' That part was easy. Jackus began to describe, as carefully and unemotionally as he could, the sequence of strange events that he had seen. The headmaster made no comment. When the piece of paper with the peculiar footprints was spread on his desk, he studied it meticulously and then signed to Jackus to continue. And he did, explaining what had happened to him and to Ann in the garage and the oddity of the thefts, with the stolen property being restored when the play was threatened. He did not forget anything, not even the disembodied fingers that had clutched at him while Marshall was tricking him with the butter muslin ghost.

When he finished, the headmaster sat silent for a moment or two, his eyebrows drawn together thoughtfully. At last he said, 'I think it would have been better if you had told me some of this before.'

'You believe me, then?' Jackus leaned forward eagerly, his elbows on the desk. 'You think I'm right?'

'I think that you have told me the truth as you see it. And that you have experienced things which are – unusual.' Mr Garner's voice was cautious. 'Whether your reasoning about them is correct, I'm not quite sure.' He tapped a hand on the desk. 'Suppose you tell me what you think I ought to do now.'

'I think you should stop the play,' Jackus said decidedly. 'Before something awful happens.'

Fleetingly, Old Garner smiled. 'That's a fairly drastic solution. I'm not sure Miss Lampeter would view it with great enthusiasm.'

'But you've *got* to. It's the only way.'

'On your unsupported word?'

'You said you believed me.' Jackus hunted desperately for something he could add. 'And it's not unsupported, anyway. You can ask Ann. She saw most of the things I saw, and she thinks – she thinks she can feel the same things the children or ghosts or whatever they are are feeling. If you'd only seen her that day in the garage –'

'Let's ask her,' Mr Garner said. He pressed a button on his desk to speak to his secretary. 'Wendy, can you send for Ann Ridley to come to my room? Straight away.'

Jackus relaxed a little in his chair. Ann would sort it out. She got so upset when she began to talk about her 'poor little children' that anyone would have to believe her. Even Old Garner.

When she arrived, she was panting, as though she had hurried along the corridor. She sat down without glancing at Jackus and he saw that her face had turned the yellowish-grey colour of her make-up yesterday. Mr Garner smiled at her.

'There's no need to look so worried, Ann. I haven't brought you here because you are in trouble. I think you may be able to help clarify a problem that Jackus here has landed me with.'

'Sir?' she said dully, not raising her eyes.

'He has told me a very strange story. Apparently, he feels that the play has raised up some kind of ghosts – or should I say poltergeists? And he told me that you would corroborate what he had said. That you yourself have had disturbing experiences.'

Ann's fists clenched on the arms of her chair. 'Why did he tell you that?' she muttered. 'What did he think you could do?'

'It is because of what he wants me to do that I have sent for you.' Old Garner looked gravely at her. 'He wants me to stop the play, and I don't feel I can do it simply on his word.'

145

Her fists clenching tighter, so that the knuckles grew white, Ann jerked her head up. 'But you can't –' In the middle of her pasty cheeks, two red spots appeared.

'You don't agree with him?'

Jackus sat very still, not quite sure what was happening. He could see that Ann was struggling for words, and he could not understand why she needed to hesitate. But when she finally spoke, he was taken completely by surprise.

'That's what he's wanted all the time,' she said fiercely. 'It's like Marshall said. He was angry because you put him in the play to punish him, so he's been doing all sorts of awful things to wreck it. And he's dreamed up this silly story about ghosts just to frighten people.'

'Ann!' Jackus yelled the word, in sheer amazement, but he could not force out another sound, and Mr Garner held up a hand to silence him. He was watching Ann curiously.

'Your story is not completely plausible,' he said in a gentle voice. 'It seems to me that things have happened which cannot all be explained in that way.'

The red spots spread, until the whole of Ann's face was scarlet. She gave a high-pitched, nervous laugh. 'Oh, that's my fault, mine and Marshall's. When we tumbled to what Jackus was doing, we decided to play a few tricks on him. To teach him a lesson. That's all.' She looked the headmaster straight in the eyes. 'It wasn't very nice, I know, but he's been *unbearable*.'

'I see.' Mr Garner looked from her heated, excited face to Jackus's pale, shocked one. 'What have you got to say to that, Jackus?'

'Don't listen to her, sir! You can't listen to her! She's lying!'

'You don't believe that you were wrong, and that she and Marshall have simply tricked you?'

'She doesn't even believe it herself.' Jackus was bellowing with rage. 'She knows about the ghosts, just as much as I do. She *knows* they're true. I can't understand why she's saying all this rubbish.'

'That seems to be the problem.' Old Garner picked up a ruler from his desk and turned it over and over in his hands. 'Here I have two of you, telling me completely contradictory stories, each of you apparently in earnest. Who am I supposed to believe? It's one against one.'

'No it isn't!' Ann said triumphantly. 'I've got someone else who'll back me up. Someone who knows all the stupid things Jackus has been doing and saying and doesn't believe them either.' She gulped for breath. 'Why don't you ask Marshall?'

Jackus knew he was beaten. He slumped down in his chair, too confused and deflated even to glare at Ann.

'Well, Jackus?' Mr Garner said. 'Shall I send for Marshall?'

Dispiritedly, Jackus shook his head. 'Don't bother,' he muttered. 'Ann's right. He'll back up everything she's said.'

'And there's no one who will support your story?'

'No sir.'

Mr Garner sighed. 'Well, then, there doesn't seem to be a great deal I can do. I will confess that I don't fully understand what has been going on. I'm not sure I believe either of you entirely. But that doesn't seem to me to be a reason for stopping the play.'

Beside him, Jackus heard Ann breathe heavily with relief. Old Garner looked sharply at her and then went on. 'Everything will go ahead as planned. But you can be certain that I shall be watching you carefully. Both of you. And I shall not expect to see any more of these – happenings.' He waited a moment, giving them a chance to speak, but no one said anything. 'Very well. You can go back to your class, then. And try to calm down before the performance tonight.'

Ann was on her feet in a second, and through the door before Jackus had stood up. He leaped after her as she ran down the corridor and caught at her sleeve.

'What was all that about? What were you playing at?' Furiously, he shook her arm. 'We could have settled it all. If you'd told the truth, Old Garner would have believed us and stopped the play.'

'I don't *want* the play stopped,' she panted. 'It must go on. It *must*.'

'But why? You know it's dangerous. You know things are getting worse and worse. And you just sat there telling all those lies. Have you gone *mad*?'

She did not attempt to answer his questions. Just shook her arm free and panted again. 'The play *must* go on.' Then she was running down the corridor again, leaving Jackus shaking with fury.

He felt hemmed in by enemies on every side. Ann. Marshall. Miss Lampeter. All the rest of the cast. They were all against him, all determined to keep him powerless. Well, let them! Deep inside him, an evil, grim satisfaction surged up. He had done his best to save them and they would not be saved. Whatever happened now, they deserved it, and he would just sit back and cheer. Tonight, in the Hall, everything that had happened so far would come to a head, like a wineglass shivering and shivering until the strain became unbearable and the thin, brittle bowl could stand it no longer. And he was *glad*.

He had just drawn level with the Hall door and, on an impulse, he pushed it open. As he stepped inside, a strong, unpleasant smell wafted towards him. Onions, vinegar, sweat. He glanced up at the stage.

The curtains were drawn across it, but instead of hanging smoothly they shook and swayed and bulged outwards, as though the stage behind were crowded with people moving about. Briefly, Jackus hesitated. Then he began to march up the Hall. *Kids messing about*, he said to himself firmly. *Need to be kicked out of here. Probably First Years*. But all the time, as he walked, he was watching the curtains jerk and shudder and his heart pounded viciously. And all the time he could smell the smell.

Just as he drew level with the front row of chairs, a hand caught at the back of one of the curtains, screwing it into a knot of creases, as though someone were trying not to fall off the stage. And a single bare foot slid out underneath the curtain.

Even through the thick streaks of dirt, Jackus could see the scar on the end of the little toe, where it ended in a jagged lump.

He stared and then, unable to believe that he was not imagining it, he took a step forward.

Immediately, he was hit by something as hard and palpable as a sheet of glass. What Ann had felt in the garage. What had been pushing at him all along, behind the falling shelves, the strange footprints, the thefts.

Misery. Misery so powerful that he would not have believed a human body could contain it. It wrenched at him, from the soles of his feet to the thin skin of his scalp, and he pressed his hands against the sides of his face to stop his mouth from splitting open into an uncontrollable scream.

It lasted no more than an instant. Even while he was trying to resist the force of it, the air vibrated in front of his eyes, like a heat haze. Without moving, the foot simply vanished as the smell evaporated and the intolerable pressure lifted.

Trembling from sheer weakness, Jackus stretched up a hand to push the curtain aside. And stood gazing stupidly across the empty stage.

'. . . nearly lost the chance. Hadn't guessed what
Jackus would do. Thank goodness they asked me.
Managed to save it. And poor Jackus. Rotten to him. But
he can't understand, can't understand. . . .

He'll get over it. He's got a life left. But not Them.
They've only got this. Their only chance. They've only
got me.

I must go back to school now.
And I know what I have to do.'

Chapter 16

The heavy red curtain swirled outwards and settled into long folds. From behind it, shuffled a stooped figure, alert and menacing, the sharp blade of the razor bright in his hand.

'They all come,' Marshall crooned, his voice carrying out over the packed Hall, 'rich and poor, old and young, they all need the barber.'

In the audience, people shifted slightly in their chairs and then settled, their attention focused on the stage.

In the dressing-room behind, Jackus could not hear the words distinctly, only the rise and fall of Marshall's voice. But his memory filled in the lines and he shuddered.

'The old devil,' murmured Stephen. 'Listen to him gloating.' Unexpectedly, he glanced at Jackus, his eyes smiling behind his glasses. 'Nervous?'

Jackus shrugged. It had been a friendly approach, but there was no time for that sort of thing now. No time for anything except listening to the voices on the stage as they came floating back through the wings. Listening and waiting.

All around him, people sat ready costumed and made up, pretending to read books, or staring at the ceiling. No one spoke. Even if Miss Lampeter had not threatened them into silence, they would not have spoken. They were tense, as though they were prepared for disaster at any moment. Mandy twisted her cross and chain between her hands while her lips moved, soundlessly repeating her lines, and Benny paced up and down, a frown on his face.

Suddenly Miss Lampeter's face peered round the door. 'Ann?' she whispered.

But Ann was already on her feet, moving heavily across the room to prepare for her first scene. Benny followed her, and the others watched them go, straining to catch the sounds from the stage.

Jackus swallowed, trying to banish the lump that was swelling in his throat, and started to think about Jarvis Williams. He needed to summon up that cheeky, carefree attitude, the confidence that made Jarvis invulnerable to the threats of Sweeney and the sinister machinations of Mrs Lovett. But it seemed impossible. Ever since that morning, when Ann had cheated him and he had seen that mutilated, impossible foot emerge from behind the curtain, he had felt nothing except a cold deadness. And he did not want to disturb that. Because he knew that, if he did, the anger would start to become unbearable. How could he possibly walk out on to the stage and be Jarvis Williams?

But when the moment came, it was unexpectedly easy. He stepped from darkness into light and there was Ann, her body padded into lumpishness and her face as grim as though she had indeed hacked with her cleaver into the flesh of human bodies. Automatically, he responded, trying to persuade her, by the sheer cheerfulness of his voice, into some warmth.

It did not work. It was not supposed to work. But as he moved round the stage he realized something that had not reached as far as the dressing-room.

The audience was enthralled. From the shadows below, there was no sound, no movement. Yet he knew that everyone's attention was directed to the small area of brightness where figures spoke and gestured, acting out the hatred and the fear that Miss Lampeter had worked so hard to build up.

When he came off the stage, Mandy, who was standing in the wings, whispered, 'It's going very well, isn't it?'

Automatically, he nodded. It was only when she had left, walking on to the stage in her turn, that he realized what exactly she had meant. It was going as Miss Lampeter had planned. The tension, the dark atmosphere, were much stronger than they had ever been at rehearsals. And yet nothing had happened. There had been no accidents, no disasters. Just the continual waiting, waiting, while the evil on the stage unfolded itself.

As the curtains swished together at the interval, there was a second of total silence and then the sound of enthusiastic clapping. Miss Lampeter pushed open the door of the dressing-room and burst in, her arms out and a wide smile on her face.

'You're all being *fantastic*. Much better than I ever dreamed. If you can keep it up for the second half, it'll be the best performance this school has ever seen.' She glanced round. 'Where's Marshall? I wanted to tell him how good he is.'

'He's up in the wings,' Benny said offhandedly. 'Just sitting.'

'Oh dear. Doesn't he want to come in and have a drink of squash with the rest of you?'

'Perhaps he's afraid we'll lynch him,' Stephen said lightly. 'He's been so good we can't bear the sight of him.'

For a moment, Miss Lampeter looked startled. Then

she grinned. 'Oh well, I suppose I wanted you to feel like that. And you can't turn it on and off like a tap. Perhaps he's better where he is. We don't want to ruin the second half, do we?'

For Heaven's sake, Jackus thought. Couldn't she see? There was a whole roomful of them, all against Marshall, and she still thought it was a game.

Even while he was thinking it, she had come across and was speaking softly to him.

'Colin, I want you to know that I'm sorry about yesterday. I was upset because everything had been going wrong. And I thought you would spoil the play. But you're being marvellous.'

'That's O.K.,' he grunted. He just wished that she would go away. Wished that the curtain would go up so that they could get on with it and get it over. Before anything happened.

Because he was aware, right down in the darkness at the bottom of his mind, that they were not safe yet. Not until the last words had been spoken and the costumes were hung up. Only then could he be sure it was all right.

And somewhere, in a hidden cranny of his brain, he did not want it to be all right.

As the second half started, he found himself drawn irresistibly towards the wings. They had all been told to wait in the dressing-room until they needed to make an entrance, but now he could not bear it. He had to be there, to watch. After a few moments, he sneaked out and hid himself in a corner where he could peer at the stage without being in the way. And, one by one, the others followed him. He could see their faces, opposite and beside him, hiding behind pieces of scenery as they gazed out at Marshall.

By the time Sweeney escaped from prison, the whole stage was ringed with eyes, invisible from the front, concentrating. Only two more scenes to go. There was Marshall's mad scene, in the cellar, with Mrs Lovett's dead body. And then the last scene, in the barber's shop, when all the good characters would finally erupt on to the stage and defeat him. That was what they were waiting for. Jackus could feel the eagerness all round him.

As the scene-shifters finished preparing the cellar, Miss Lampeter came swooping across the stage with the cut-throat razor in her hand. She placed it neatly on the seat of the barber's chair, ready to be carried on for the last scene. Then she glanced round irritably.

'Where's Ann? She should be on stage by now. We can't put the curtain up until she's there. Someone go and get her.'

'I'll go,' Jackus said. But before he could move, Ann was there, walking woodenly past him as if she were asleep. She paused for a moment at his side and he thought that she was going to speak, but the next moment she moved on, walking carefully on to the stage and lying down in the centre, arranging herself as if she were dead.

Always before, she had sprawled out, her arms flung wide and her face staring up at the ceiling. But this time she lay hunched and curled, with one arm coiled underneath her and her back to the audience. Jackus just had time to wonder why she had changed it before the curtains opened and Sweeney, mad and desperate, darted on to the stage.

He drew everyone's eyes. In the terrible quietness of his insanity, he seemed more powerful, more menacing, than ever before. Jackus found that he could not look away from that contorted, cruel face that seemed to radiate hatred out across the dark Hall. It was so strong that he started to shake and, feeling his knees tremble, he sat down hastily on the barber's chair beside him.

'I shall not be defeated!' Marshall was hissing. 'Not by all the plans that they can lay against me. I am Sweeney Todd, and I am stronger than the law, stronger than goodness – stronger than death itself!'

He spun round, beginning to hunt for his lost jewels, and Jackus was so absorbed that he did not realize for a few minutes that something was wrong. Not on the stage, but here in the wings, where he was sitting. The oddness niggled away at a corner of his brain, demanding attention. As Marshall prowled towards the back of the stage it suddenly broke the surface of Jackus's consciousness.

There was something missing. Something that was not where it should be. He frowned, glancing away from the stage, and all of a sudden it came to him.

When he sat down on the chair, there had been nothing on the seat. But there should have been. Miss Lampeter had put the razor there, just before she asked where Ann was.

Fear squeezed at Jackus's chest. He slid off the chair and started to feel around on the ground, telling himself that the razor must have slipped off. But even while his hands were searching, he knew that he would not find it. He remembered, too clearly, how Ann had paused beside the chair, and he knew now why she had changed her position on the stage. Why she had been so determined that the play should go on. The razor was out there on the stage, hidden in the thick folds of her skirt. Where no one would see it, until it was too late.

Straightening, he looked out over the stage again. Marshall had finished his useless search for the jewels and was moving towards the body slumped in the middle of the stage.

'Why, poor Mrs Lovett,' he crooned, 'I do believe you are dead.'

He stooped forward, peering, and Jackus saw Ann move, almost imperceptibly. Her eyes opened, facing away from the audience, and her lips parted. Everyone else was watching Marshall, unaware.

All around Jackus, the air trembled, as if he were in the bowl of a great, vibrating goblet of glass. For the second time that day, his skin seemed to suck up feelings from outside, feelings that were not his own, but that were so powerful he tingled.

Only, this time, it was not misery he felt. It was joy. An expectant joy, that any moment would release itself into triumph. Marshall – Sweeney Todd – the old man – hung suspended by a thread above an abyss of retribution that would make up for all the pain, all the betrayal.

Yes, said Jackus's brain. *Yes, he deserves it. Yes!*

Marshall stooped lower, his neck stretching out pale and vulnerable. As Ann began to sit up, something deep down, deeper than his brain, deeper even than his feelings, set Jackus moving. Because he could not let it happen after all. Not even to Marshall. Or especially not to Marshall.

Launching himself forwards, through air that seemed to resist and cling, he flung his body at Ann across the wide space of boards that separated them, trying to reach her before her arm emerged completely from the folds of her skirt.

He was almost too late. Disjointedly, he saw the blade of the razor gleam, slicing upwards at that pale, exposed neck. Saw Marshall's eyes widen as he realized, incredulously, what was about to happen. Then his body crashed at full speed into Ann's moving arm, knocking her sideways with a crunch. Taken by surprise, she started to yell almost immediately, struggling to free herself.

'You've got to let me! I must! He's evil and They've waited so long!'

Desperately, Jackus tried to keep a grip on her, but she was heavier than he was and he felt his hands begin to slip. Completely unaware of voices shouting and feet running all around, he hunted despairingly for something else that he could do, some way he could stop her before she managed to wrench her right arm free and jab it up towards Marshall's pale, terrified face.

And, magically, it came to him. Flinging his head

back, he yelled, not at her, but at the humming, crowded air about him:

'If you go on hating the old man, he'll live for ever! Marshall doesn't matter. It's *you* keeping him alive. And you don't have to. You're free! You're dead!'

Ann's body suddenly went slack in his hands and, falling sideways on top of her, he heard, incredibly but distinctly, the sound of hundreds of feet running. Not desperately, as if they were pursued, but joyfully, skipping, their bare soles pattering lightly on the ground, like children let out of school at long last.

Marshall gave one shocking sob and dropped to his knees.

The next moment, people were all around them, angry voices sounding in their ears.

'What do you think you're doing?' Jackus heard Miss Lampeter say, as if she were a long way off. And, beside her, Old Garner's voice rumbled as he bent to feel Ann's arm. 'I'm very much afraid the bone is broken.'

The only thing that was clear was Ann's face as she turned to Jackus, her cheeks pink, her eyes full of surprise and delight.

'Did you hear? Oh, Jackus, did you *hear*?'

Marshall was down by the canal, his tall figure no more than a shadow in the darkness, his head turned to look up the path as he waited. Jackus swung his bag against his legs and marched up to him, his ears still glowing from the pleasant things Old Garner had said while the chaos was being sorted out. He felt that he could deal with anything now. Even Marshall.

And as he drew level, it was Marshall's eyes that shifted, Marshall who looked away in embarrassment and turned his head awkwardly.

'Hi, Colin.'

'Hi, Colin,' Jackus said.

'I thought I ought to say something.'

'You don't have to.' Jackus grinned at him. 'I don't think I'll believe you if you start grovelling.'

Marshall smiled wryly. 'Wouldn't hurt me though, would it? Because you did save my life.'

'Aw, shucks,' Jackus drawled, mock-American, 'I'd have done the same for anyone.'

'Even me. Even after everything.' Marshall raised his eyebrows. 'That's hard to take, you know. Humble pie and all that. Not really my sort of thing.'

'I expect you'll manage to gulp it down.'

'It would go down a bit easier if I could do something for you. To make up. Would you like –' Marshall hesitated '– would you like me to go and see Old Garner tomorrow? Explain to him that taking those tape recorders was all my idea? That I dared you to do it, and I was there all the time?'

'No,' Jackus said. 'It doesn't matter any more.'

'But there must be something I can do.'

Jackus looked up at him. The clever face was shadowed, but even in the dimness he could see that it was apparently earnest and sincere. For the time being.

'No,' he said slowly, 'I don't think there's anything at all you can do for me, Marshall. Not any more.'

He walked past, along the tow-path by the black water, and as he went he felt suddenly light, as though for the first time in his life there was nothing dragging after him.